Penguins: *The Animal Answer Guide*

Penguins

The Animal Answer Guide

Gerald L. Kooyman and Wayne Lynch

The Johns Hopkins University Press Baltimore

Printed in the United States of America on acid-free paper
9 8 7 6 5 4 3 2 1

The Johns Hopkins University Press
2715 North Charles Street
Baltimore, Maryland 21218-4363
www.press.jhu.edu

Library of Congress Cataloging-in-Publication Data

Kooyman, Gerald L.
Penguins : the animal answer guide / Gerald L. Kooyman and Wayne Lynch.
pages cm
Includes bibliographical references and index.
ISBN-13: 978-1-4214-1050-0 (hardcover : alk. paper)
ISBN-13: 978-1-4214-1051-7 (pbk. : alk. paper)
ISBN-13: 978-1-4214-1052-4 (electronic)
ISBN-10: 1-4214-1050-8 (hardcover : alk. paper)
[etc.]
1. Penguins—Miscellanea. I. Lynch, Wayne. II. Title.
QL696.S473K66 2013
598.47—dc23 2012050455

A catalog record for this book is available from the British Library.

Unless otherwise noted, all photographs in this book are by Wayne Lynch.

Special discounts are available for bulk purchases of this book. For more information, please contact Special Sales at 410-516-6936 or specialsales@press.jhu.edu.

The Johns Hopkins University Press uses environmentally friendly book materials, including recycled text paper that is composed of at least 30 percent post-consumer waste, whenever possible.

I have often had the impression that, to penguins, man is just another penguin—different, less predictable, occasionally violent, but tolerable company when he sits still and minds his own business.

Bernard Stonehouse

Contents

Acknowledgments

Many colleagues have clarified questions I have had about the biology of penguins. Paul Ponganis and his research team of Jessica Meir, now an astronaut candidate in training in Houston; Cassondra Williams, now at the University of California–Irvine; and Birgitte McDonald, a postdoctoral researcher at Institute of Bioscience Aarhus University, Denmark, all helped with the initial proposal and later with aspects of penguin physiology and biology. My own former graduate student Jefferson Hinke, now at the Antarctic Ecosystem Research Division, was always very helpful, both while I was reviewing his thesis and in discussions about the conditions at the Antarctic Peninsula.

Outside of our penguin enclave at Scripps Institution of Oceanography, I have some exceptional colleagues who responded quickly to my questions. I'll mention only a few: Yvon LeMaho, Université de Strasbourg, with his students, conducted some of the most insightful research on king and emperor penguins of any group. Phil Trathan, head of conservation biology, British Antarctic Survey, took time to explain the status of penguin populations and the reliability of the estimates, especially of macaroni penguins around South Georgia Island. Julie Hagelin, Institute of Arctic Biology, University of Alaska–Fairbanks, who knows penguin feathers well, helped me discover just how remarkable they are. Linda Henry, supervisor of birds, Sea World Adventure Park, whose workplace is a near neighbor to mine, is always available for questions about many aspects of husbandry and the Encounter's population. She is a walking encyclopedia about the birds under her stewardship and knows every individual by name. It is a great asset to be able to see for myself some aspect of any one of several species of penguins on exhibit there.

In the course of this writing project, I discovered penguin paleontology. I am grateful for the reports of ancient penguin species and for the permission of Dan Ksepka, North Carolina State University, to use a figure scaling the giant penguins to modern penguins. Having acknowledged so many in this endeavor, I qualify that all things said in the book and all interpretations are mine.

There is always the matter of funding for travel to the remote areas where penguins live. I have been lucky to work in some of the most remote areas, where I could not have gone without the support of the National Science Foundation Office of Polar Programs, under the direction of successive program managers Richard Williams, Polly Penhale, and Roberta

Marinelli. No one could ask for more cooperative and supportive administrators.

Most important of all for the completion of this book was the help of Melba, my wife, who read every word and offered countless suggestions in her effort to keep my syntax straight and my answers clear. We both did the book for the grandkids, ours (Hannah, Zack, Sophia, Abby, Lili, Elle, and Olivia) and everyone else's.

Introduction

In this book I address some basic questions: (1) What is a penguin? (2) What is the origin of penguins? (3) Where are they are found now, and where did they live in the past? and (4) What adaptations make it possible for them to live where they do? After 30 years of working with penguins, I thought I had learned a lot, but over the past two years of discovery of all penguins, my learning curve has taken a steep increase. It has been time well spent, and I hope readers of this book will pick up that sense.

Hemingway once wrote, "I always try to write on the principle of the iceberg. There is seven-eighths of it underwater for every part that shows." I like the iceberg metaphor; it describes my goal that my answers will provide a starting point and an aid for the reader to pursue in more detail the many knowns and unknowns of this incredible group of vertebrates. It has not been an easy experience trying to translate my thoughts into words on a page. Like John Muir, I find that "writing is like a glacier, one interminable grind." Fortunately, my grind has been interrupted with informative trips back to Antarctica. In the course of those travels, I have seen emperor and Adélie penguins in a different way. How do I as a writer express what I have seen to a broader audience? I have to think more historically. I have learned about the people who were most likely the first to see (and probably hunt) penguins, as well as the first Europeans to report observations of penguins. I wonder about the hardships suffered by Bartolomeu Dias and his crew, who were the first to sail thousands of miles to the Cape of Good Hope in 1488 and see African penguins. I wonder who, in recent times, the first person was to see all 17 species and how many years it took. Now, thanks to NASA images obtained from space satellites, we can view the remotest of all penguins via the Internet and see every known emperor penguin colony.

Beyond their biology of breeding, penguins have wonderful adaptations that are essential to existing in the marine environment. Penguins are the stuff of nature's miracles. To mention a few, they have the best body-form design for air breathing vertebrates (animals with a backbone) and the best "dry" suit for diving, and they evolved drag reduction by "air lubrication" millions of years before flight engineers invented it in recent times.

Those of us who have worked in the ice fields of Antarctica enjoy and value the most pristine environments now existing on the planet. Because of these travels we are highly sensitive to changes that are occurring elsewhere, and this is the dark side of my view from the perspective of the penguin and of a field biologist. We once had a much simpler life. Are we a victim of "Dollo's principle," which says we can never go back to those simpler times? It seems that we are determined to squeeze the last molecule of fossil fuel out of the planet and extract the last edible fish from the ocean. These practices are not without cost. Penguins are sentinels reflecting unsustainable activities, and their numbers are decreasing. The one exception is the king penguin. In the face of resource extraction, oil spills, and global warming, its population is growing. Why is that? We need to learn more about the success story of this species. King penguins give me hope. My greatest pleasure is learning more about the wildlife of this planet, and I know there is a solution to our problems; part of it is to live simply and sustainably.

Penguins: *The Animal Answer Guide*

Chapter 1

Introducing Penguins

What are penguins?

Penguins are flightless birds that spend most of their lives at sea. For a few weeks each year the breeding birds devote some time to incubating their egg or brooding the chick. They are exclusively found in the Southern Hemisphere, unless you count the few Galápagos penguins that just barely slip over the equator from time to time. They are not limited to cold climates on land, but the waters where they hunt and feed are cold or frigid. Having evolved for their aquatic habitat, their wings are flippers and their black-and-white coloration provides protective counter-shading in the water.

Most birds have a somewhat upright, leggy stance, in which the legs are a very prominent feature of the whole body. Penguins, however, with most of the leg hidden in the body and feathers, have shorter legs than most birds. The penguin stance is more upright than that of other birds because the femur relative to the backbone extends from the hip at an angle of about 120 degrees, compared with the 60 degree orientation of other birds. The result is that penguins when standing have an upright posture, humanlike, rather than the horizontal alignment of the body axis of most standing birds. Also, the penguin's tarsometatarsus, a bone found only in birds, is extremely short. It is the leg bone just above the foot, and the equivalent to the tarsal and metatarsal bones of the human foot. The shortness of the penguin leg, resulting from the size of its tarsometatarsus bone, is a most useful skeletal feature to paleontologists (scientists who study fossilized animals) in identifying a fossilized bird as a penguin.

There are three basic gaits of penguins while traveling on land: (1) a waddling or side-to-side swaying walk, as in the emperor penguin; (2) running with the body angled forward, as the little penguin does; and (3) hopping, which is very useful on boulder-strewn shorelines and cliff faces; an example is the eponymous rockhopper penguin. While traveling upright, all penguins except the emperor penguin hold their wings extended. The emperor penguin keeps its wings against its sides. Despite the short legs and awkward appearance, the energy retained in a stride, known as the recovery rate, is better for penguins than for humans. Polar penguins, the ones that travel over snow, all have a fourth form of overland travel called "tobogganing." The birds flop down on their stomachs and push forward alternately with their forward-reaching three-toed, clawed feet. They probably steer by pushing harder with one foot or the other. The wings add stability to the body when it tends to roll from side to side. Because it spreads the bird's weight over a large area of the snow, this is the most efficient and fastest form of over-snow travel. However, tobogganing causes feather wear, which in turn may reduce swimming efficiency. There is no free ride in nature, but tobogganing can be a time for a little napping if it is done on an extensive plain of ice. Once into the routine, a bird may rhythmically cruise along, eyes closed, and come up with a start if it bumps into something unexpected, such as a person in its path. At high speed, a tobogganing bird adds extra thrust by striking its wings against the snow vigorously and in unison. Such travel can be as fast as a man's trotting speed.

In water a penguin's feet are directed backward and in line with the flow of water from the front of the body. This reduces body drag while swimming. The bottoms of the feet may be directed upward when cruising. When the penguin is maneuvering a turn, the feet form a rudder in line with the tail. Slowing down is aided by pointing the tops of the feet downward to act as a brake. Penguins' large feet have three major toes and a small vestigial toe (a toe with no apparent function), and their feet are slightly webbed.

The wings are shaped like blades and covered by very small, scale-like feathers. There are six major bones in the wing: the humerus, the radius and ulna, a fused carpometacarpus, and three phalanx bones, two between the carpometacarpus and a single bone at the tip. The bladed wings act as hydrofoils; propulsion is provided by lift from the wing during both the upthrust and the downthrust. They are not paddles. During hydro-flight, or underwater swimming, the body is held rigid by the stiff spine, and the neck is retracted except when striking at prey. The form of the rigid, perfectly spindle-shaped body is ideal for low-drag swimming. Likely it is one of the lowest-drag designs of an aquatic animal, except for the tuna.

Additional hydrodynamic properties are the feathers. I have been asked at times whether penguins have feathers or fur, because their feathers,

Tobogganing emperor penguins.

which are highly modified, or altered from the basic form, look superficially like fur on the outside. The feathers of penguins, unlike those of most or all other birds, are uniformly arranged over the body and are not in feather tracks. All penguin feathers are short, and there are four main types: (1) the body or contour feathers, (2) the wing feathers or remiges, (3) the tail feathers or rectrices, and (4) adult down feathers. The wing feathers, which superficially look like scales, are perhaps the most struc-

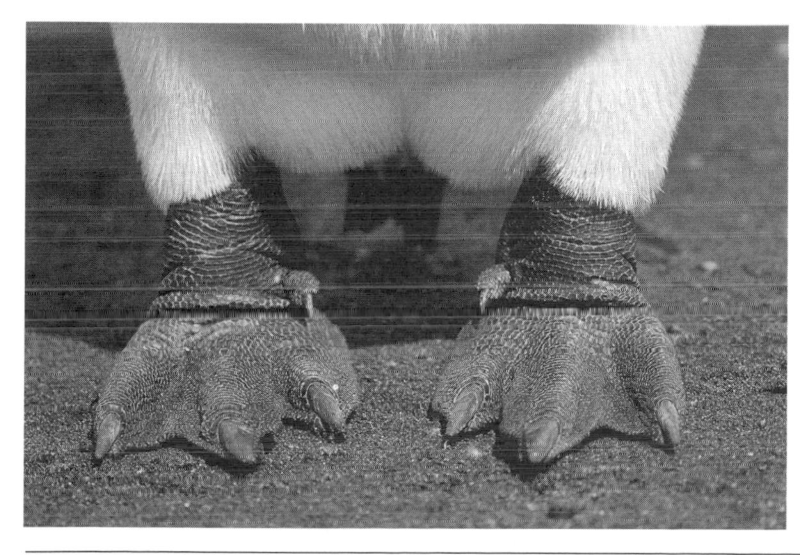

Webbed foot of a king penguin.

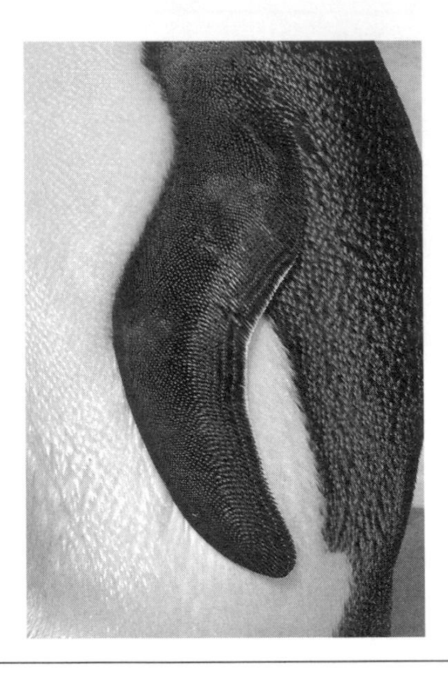

Emperor penguin wing (flipper).

turally unusual among birds' feathers. That may be why the oldest extant (currently existing) genus of penguins, which includes the king and emperor penguin, was named *Aptenodytes*, which means "without feathers" or "featherless diver." The tail feathers, or rectrices, have a round section at the base, the calamus or quill, which is about 35% of the total length. Where the pennaceous vane, or the main body of the feather, begins, the main shaft, or rachis, has a deep groove that extends the rest of the length of the feathers. The vanes are short and soft.

The extremely rigid, shortened, and strong forelimb (the wing) is a fine example of how the typical vertebrate body can be modified for a special task, in this case propulsion. The wings serve as flippers, and penguins are the only birds so equipped. There is almost no flexibility in the bladelike wing, in contrast to the wings of flying birds that also dive and use their wings for propulsion; those birds shorten their wings when diving, by partially folding them. In addition, when flying species are at rest, they fold their wings against their sides. Think how awkward an albatross would be if it could not fold its wings when resting at the nest or on the sea surface. All of a penguin's wing feathers are highly modified. The leading-edge covert feathers of the wing are short, scale-like in appearance, and the same length on both sides of the rachis. They are evenly distributed on both the front and back surfaces of the wing, densely packed, and grade smoothly across the wing into the secondary feathers at the trailing edge. Toward the tip of the wings, the primary feathers appear to be nearly the same as the secondary feathers.

The emperor penguin's body feathers can serve as the extreme model, because it inhabits the coldest environment of any penguin. Each contour feather of the emperor penguin is divided into distinct regions along its flattened shaft. The plumaceous (downy) region begins at the base of the feather, closest to the body, and makes up about half of the 5 cm feather length. By comparison, a typical land bird, such as the Coopers hawk, has a downy region on half of its 10 cm contour feather. Beyond the plumaceous segment and up to the tip, the feather is pennaceous, with barbs branching from the rachis. The barbs are of uniform length on either side of the shaft and have loosely connected barbules, or proximal and distal branches, that grade to disconnected hairlike strands near the end of the barb. The tip of the feather is without barbules toward the end of the barb. In the barbule-free part, the barbs lengthen and bear the surface color, white on the breast and black on the back (or some other color for penguin species with additional colors). Near the shaft base, there is a separate inner plumaceous extension, or hypopenna, also called the afterfeather; the afterfeathers serve as an insulating layer of the feather coat. Perhaps the greatest insulation comes from the numerous downy feathers that have no central rachis and are clustered around the main contour feather. These shorter feathers add much to the volume of the undercoat. Besides their insulation properties, the contour feathers may act as flow dampeners on the body and reduce drag, as has been suggested for the rubberlike skin of dolphins and the rough skin denticles (pointed projections) of sharks. (For more about feather function to reduce drag, see chapter 5.)

Feathers are different on different parts of the body. The breast feathers are the most similar to each other and form a tight overlay. There is a gradation of feather length toward and on the head. The longest feathers are the tail feathers at 11 cm in the emperor penguin and 9 to 11 cm in the Adélie penguin. The strong central shafts of the tail feathers are black, as are the long, symmetrical barbs branching from the shaft.

Birds do not have teeth, so the bill is highly variable depending on how it is used. All penguins feed on fish, krill, or squid, which are found mainly in the water column rather than on the bottom, and the bill is shaped according to diet. The bill is composed of keratin (a substance present in fingernails), which overlays the bones of the upper and lower jaw. These bones form the basic shape of the bill, and the keratin adds some further features. The heavy bills of the *Eudyptes* group are the most unusual. They are thick and colorful, usually red or orange. Then there are the "plain Jane" bills of most other penguins, which are gray to black; some are moderately heavy, as in the banded penguins, or tubular and small, as in the little penguin and the brush-tailed penguins. The banded penguins have a hook on the end of the beak that fits into a recess in the lower part of the bill and probably helps in grasping fish firmly. In the brush-tails, the gentoo stands apart with a bright orange bill. Enclosed within the bill of

Rockhopper penguin tongue and palate.

all penguins are the serrated tongue and palate, which aid in gripping and swallowing prey. The projections are directed down the throat to prevent wriggling prey from escaping.

The nares (nostrils) for breathing are part of the bill and are located at the junction between the feathers and the base of the bill. There is also a groove on each side of the bill down which flows the fluid from the salt gland, located above the eye in a depression of the skull. This concentrated saline solution drips off the end of the bill.

The most distinctive features of the head and skull of the penguin besides the bill are the large, linear depressions in the top of the skull above the eye socket. These are the pits that accommodate the salt gland. The eye sockets are the largest spaces in the skull, and the two eyes occupy more space than the brain does. Penguins' eyes range in color from dark brown or black to bright red and, of course, yellow in the yellow-eyed penguin.

The capstone of the penguin image of robustness is the underlying feature of the breast muscles. The size and shape of those pectoral muscles dominate the shape of the body, as they do in many other birds; but even among birds, the penguin breast is exceptional. By weight, the two muscles, supracoracoideus and coracoideus, represent more than 75% of the body mass and are the largest single feature of the body. Few vertebrates have such a dominating feature, but unlike the tusks of elephants, the legs of ostriches, or the long jaws of crocodiles, the breast muscles are hidden under the feathers and within the smooth shape of the body.

Why are penguins important?

People are charmed and enchanted by penguins, and as a result a tourist industry centers around them. Currently some 40,000 tourists a year pay

Tourist with king penguins.

considerable sums of money to voyage to the sub-Antarctic and beyond to Antarctica. Countless others go to South America, South Africa, New Zealand, and Australia in part to see penguins. Near Melbourne, Phillip Island alone has over a million visitors per year to the little penguin sanctuary. Many—some would say most—of these tourists are trying to satisfy their desire to see penguins in the wild. Perhaps that interest is due to some of the traits we share with penguins: they walk upright, they are gregarious and noisy, and (if we approach their nests) they can be aggressive. Some of their prenuptial behavior even suggests affection between cuddling couples. Penguins are also beautiful animals that give us peace of mind when in their presence. Ultimately, they are approachable; it is possible to stand near or among penguins, and, when this is done slowly and quietly, they appear not to be disturbed. These qualities make them excellent ambassadors for wildlife. Their charisma offers us the most recognizable icon of Antarctica, even though most species live elsewhere.

Penguins were an important source of survival food for early Antarctic and sub-Antarctic explorers. Since they are fish-eating birds, however, one suspects that penguin meat has an oily, fishy taste that is not especially desirable unless a taste is acquired. Penguin eggs were once sought after, and to this day in some places they are still collected as a food supplement. In an earlier time the king penguin also served as a fuel source.

From a medical perspective, penguins have certain physiological attributes that are worth understanding. They routinely dive and remain under water for long durations, with the result that the oxygen levels in their blood and tissues decline to remarkably low levels near the end of some dives. Indeed, the levels often go below what can be tolerated by humans and other terrestrial animals. For medical scientists there is more than idle curiosity in understanding this tolerance to hypoxia, or oxygen deficiency. What mechanisms underlie penguins' management of stored oxygen? Would uncovering these mechanisms and understanding how their bodies can withstand low oxygen levels provide us with medical treatments? For example, studying penguin tolerance to hypoxia might help us develop better methods of harvesting and preserving organs for transplant. Such research might also lead to new treatments for heart attacks and strokes. During heart attacks and strokes, tissues and blood experience dangerously low oxygen levels, but these same low oxygen levels seem not to faze diving penguins. Perhaps one day we will discover a substance in penguins that will lead to a new class of drugs that can be used for emergency treatment of humans when the circulatory system fails.

Penguins are also useful to engineers. There is always the inevitable question of how penguins withstand the cold and even near-freezing waters in which they forage. The most extreme are Antarctic waters, where the seawater hovers around 28.4°F (–2°C). What is it in the design of their feathered coats that keeps them warm in such cold seas, and is it worthwhile to try to build a fabric that mimics the properties of penguin coats? There are other aspects of how penguins' bodies work that are worth studying, but these few points illustrate the wealth of special adaptations penguins possess in order to survive in their extreme environment.

Finally, as a high-profile group, penguins are ideal as sentinels for the environment. They are universally identified. Everyone knows what a penguin looks like. They are large for birds and effortlessly recognized on land and at sea. They can be counted and observed with little difficulty in some places. Generally we know more about penguins than about our own common backyard birds. As large predators, they not only have a strong influence on the food chain but also provide a good indication of the abundance and distribution of their food base. Two basic questions are immediately raised when an increase or decline occurs in their populations. Is their foraging habitat changing, or is the breeding territory changing, or maybe both? For most penguins, those that live in warm climates, populations are decreasing, and we seek answers to those two basic questions: Is this because of a decline in breeding habitat on land or in food availability at sea? For most species it is usually both, and these confounding factors make it difficult to pinpoint a major cause. There is disturbance of or encroachment on breeding sites by humans or by other species, often by species introduced by humans. At the same time, there are often competitive fisheries nearby that reduce the prey of penguins. Unlike their northern relatives, penguins in the Antarctic suffer little from human intrusion into the breeding habitat. However, fisheries are always searching and expanding to new regions, Antarctic waters are experiencing a negative effect on their food base as exploitation of krill and fish increases in the Southern Ocean. Indeed, krill, a major food base, has declined by about 50% in the Antarctic Peninsula, one of the richest krill-producing areas in the world. Nevertheless, perhaps the most consequential cause of decline in penguin populations is the result of climate change, because it has a strong, usually negative effect on habitat and the food base. The worldwide increase of greenhouse gases, largely emanating from the Northern Hemisphere, influences Southern Ocean habitats. The warming occurring in the polar regions has decreased the extent of sea ice, disintegrated ice shelves, increased snowfall, and caused changes in wind and water current patterns. The combination of these factors illustrates why such highly observable species as penguins make them ideal biological monitors, sentinels, or indicators of environmental change and what the future may hold for all of us.

Why should people care about penguins?

One of the first questions we ask about almost any item of nature, whether astronomical, biological, or chemical, is "What is it good for?" It is also one of the most difficult to answer. This is a basic concept of our own self-interest, and only in recent times have we had the luxury to go beyond the self-interest query. The answer to the question depends upon who is asking: a child, a corporate CEO, a conservationist, a creationist? a person in some other category? Corporate economists, conservationists, and marine biologists have similar concerns but different objectives. For those groups, penguins are important because they are near the apex of marine predators. In their ecosystem, where top-down forces play a large role, they are often one of the keystone species, at least among species that are abundant. They are also important and very visible indicators of how stable particular ecosystems are. Examples of large populations of penguins that have substantial effects on the food web are the sub-Antarctic and Antarctic species, some of which are the most abundant avian species in the world. The same was once true of the cold-temperate species, but most of them are now in decline as a result of competition with fishers, chronic oil spills, and other degradations. In short, it is degradation of habitat resulting from human corporate activities. This in itself is competitive with one of the largest of industries, entertainment or tourism. Many of the wild marine areas of the Southern Hemisphere are popular with tourists because of the abundance of penguins.

For the child or the ethicist, the importance of penguins is not just as a source of travel entertainment but also in the image of the penguin that they see in toys, movies, and clothes. The penguin becomes identified with pleasurable images or experiences, and this creates a concern for penguins as a group. This association increases people's interest in preserving the habitat and leads to a more intimate regard for some of the greatest wild treasures of the planet. It prompts a level of concern higher than interest in a product that can be marketed, a concept that is vital to preserving the health of the diverse marine habitats around the Southern Hemisphere.

As for the creationist, if penguins are a product of intelligent design, and Genesis 1:26–28 is taken seriously, then penguins must be a group worth caring about and preserving because they are part of creation. It is reasonable to assume that God does not like what he creates to be abused by careless or profligate activities of humans, no matter how embedded the command to "subdue" the earth "and have dominion over the fish of the sea, and over the fowl of the air, and over every living thing that moveth upon the earth" is in the way one perceives wildlife.

How many kinds of penguins are there?

Traditionally there are 17 species of penguins, although some authorities have argued that 3 species, the little blue, yellow-eyed, and rockhopper penguins should be further divided into additional species and that the royal and the macaroni should be considered a single species. Some say it is a "hot debate" whether the macaroni and royal penguins should be lumped together. Despite that view, L. Spencer Davis and M. Renner, in their book *Penguins*, present several diagrams showing the two as separate species, but with the royal in single quotes. In the most current diagram, they are presented as separate species. If the "splitting" position carries the day for these two *Eudyptes*, and if other species are split, the total number of penguin species could increase to about 20, but for now it's probably prudent to assume 17.

What is the current classification of penguins?

Those 17 species are grouped or classified into six categories, or genera. This grouping is a means of managing the complexity of the entire group; we begin with the broadest terms for penguins and become more specific with each lower group in the hierarchy. At the final level we use a binomial nomenclature, developed by Carl Linnaeus. After his first *Systema Naturae* was published in 1735, many later editions appeared; in 1758, with the tenth edition, the zoological nomenclature was added. To reach this level, the history of penguins was reconstructed by a stepwise process from the fossil record, the age of the fossils, and their changing morphology (form and structure) to the current family of penguins. The process involves constructing relationships from synapomorphies, or shared derived traits; not only morphological traits, but also molecular and behavioral characteristics are included as penguins are grouped with the nearest relative that has traits in common with penguins. The end result is a group of cladograms (branching tree–like diagrams) that illustrate not only the relationship of penguins to each other but also the possible origin of penguins from the most closely associated birds from the more distant past. Since the cladograms are based on a variety of data and assumptions, with many gaps in the fossil record, it is not surprising that there are many renditions of when penguins evolved. A general agreement is that the closest relatives to the penguin group are flying birds from the Procellariiformes order (petrels and albatrosses) and the Gaviiformes order (grebes and loons), and that the Sphenisciformes order (penguins) was established over 60 million years ago (ma). This occurred near the time of the Cretaceous-Tertiary (Cenozoic) boundary. This boundary signifies the cataclysmic event of a giant meteor impact that resulted in mass extinction of many species of plants and animals, including the dinosaurs. Some birds survived, including, it is believed, ancient penguins. At present the number of extant (still living) species is 17 and the number of extinct species at last count was 26; their relative size ranges from the little penguin to an extinct species about twice as tall as the emperor penguin; the earliest fossil species lived 60 ma; and the last known extinction of a species was nearly four ma.

The present classification of penguins was developed using methods that range from the time-honored and earliest method of comparing morphological characteristics of fossils and extant species, to analyzing molecular phylogenies (DNA sequencing) of extant species. Here is one version of the classification of present-day species based on morphological characteristics.

Order Sphenisciformes
- Family Spheniscidae
  - Genus *Aptenodytes*
    - Species *forsteri, patagonicus*
      - Common names: emperor, king
  - Genus *Pygoscelis*
    - Species *antarctica, papua, adélie*
      - Common names: chinstrap, gentoo, Adélie
  - Genus *Eudyptula*
    - Species *minor*
      - Common name: little

Genus *Spheniscus*

Species *demersus, magellanicus, humboldti, mendiculus*

Common names: African, Magellanic, Humboldt, Galápagos

Genus *Megadyptes*

Species *antipodes*

Common name: yellow-eyed

Genus *Eudyptes*

Species *pachyrhynchus, robustus, sclateri, chrysocome, chrysolophus, schlegeli*

Common names: fiordland, Snares, erect crested, rockhopper, macaroni, royal

Any classification of penguins must take note of the influence of continental drift. One of the earliest summaries based on the information at the time was B. Stonehouse's report that penguins evolved in tropical to subtropical seas of the late Cretaceous to the early Miocene period. The central area was a large arcing island continent called Tasmantis (Zealandia), which was a land connection that reached from present-day Campbell Island, New Zealand, to New Caledonia. It altered between a single landmass and an archipelago through the late Cretaceous to the Miocene. Tasmantis was a part of Gondwanaland until it separated from Antarctica; it then broke off from Australia in the late Cretaceous to the Paleocene. When these separations occurred and the subcontinent divided into a series of islands, the currents changed and turned cold around southern New Zealand. It was under these cooler conditions that the present-day radiation, or diversification, of the current species of penguins occurred, and for the second time in the Cenozoic period, Antarctica became ice-covered. The greatest diversity that developed was in the crested penguins of New Zealand and its nearby islands, as well as most other sub-Antarctic islands.

After studying fossils and present-day species, scientists generally agree on the approximate origin age of the various genera. They have concluded that the most likely hypothesis is a center of origin associated with Gondwanaland when Antarctica was still attached to Australia and South America and when New Zealand was relatively close to the supercontinent. The timing of the common ancestry of extant penguins is still uncertain. The divergence occurs between the basal lineage of *Aptenodytes* at about 12 ma, based on the traditional paleontological aging processes, versus a report that it diversified about 40 ma, based on molecular assumptions. Noteworthy is that only the emperor and Adélie penguin have been able to move far south into Antarctica's cold zone with the cooling of the planet during the Cenozoic, which process has been aptly labeled "from greenhouse to icehouse." To summarize, by the beginning of the Cenozoic, the continents of

South America, Australia, and Antarctica were separated, but the southern continent had not reached the South Pole, and New Zealand had moved away from Antarctica. The drift trend continued, and by the late Miocene the Spheniscidae was established and the modern penguins spread out, but to a lesser degree than the earlier ones had, and all were restricted to the Southern Hemisphere and cold temperate to polar environments.

What characterizes the major groups of penguins?

The group of penguins thought to be the oldest, the king and emperor penguins (*Aptenodytes*), are also the largest penguins. They range from about 95 cm to 115 cm, and 15 kg to 25 kg, respectively. Both have bright yellow auricular (ear) patches and long and delicate bills with softly red to pink mandibles (lower jaws). Their feet are long and black.

The three species of brush-tailed or *Pygoscelis* penguins, so called because of their rather coarse, stiff tail feathers, are the second-oldest group, and the gentoo is the third-largest penguin at about 71 cm and 6 kg. The gentoo is also the most colorful, with a bright orange bill and feet. The Adélie penguin is the least colorful, with all black-and-white feathers and pink feet. The chinstrap penguin is of similar size and weight to the Adélie penguin, and both of them are slightly smaller than the gentoo. The chinstrap has a distinctive black straplike line running from side to side of the head and under the chin.

The medium-sized crested penguins (*Eudyptes*) have yellow crests (feathers on their foreheads that look like a crest), which are distinctive among the six species. All have short and heavy red bills and red eyes, except for the dark eye of the erect crested. All have pink feet. They are 55 to 70 cm in length and weigh 2.5 to 4.5 kg. The royal penguin, which, as stated earlier, some believe is only a color variation of the macaroni penguin, is the largest, and the rockhopper penguin is the smallest of the group. There are also disputes about whether the rockhopper is a single species, with some believing it should be split into a northern and a southern species.

The yellow-eyed penguin is a single species within the genus *Megadyptes*. However, some believe it should be split into two species. It has yellow eyes! There is also a yellow band of feathers extending from eye to eye across the back of the head. The other single species genus is *Eudyptula*, it contains the little penguin. A blue- to gray-backed penguin with a white breast and pale yellow eyes, it is the smallest of all penguins at 40 cm and 1 kg. It is suggested that possibly a second species, the white-flippered penguin, should be designated in this genus.

There are four species of banded penguins in the genus *Spheniscus*—Galápagos, Humboldt, African, and Magellanic, which can be distinguished by

Young emperor penguins in a loose grouping.

the amount of banding (ordered here from least banded to most banded). There are two body bands, one of them a white line that runs down the body (vertically when standing) and terminates at the top of the legs. The second band outlines the cheek area of the head. The smallest and least distinctively banded penguin is the Galápagos. It is 50 cm and weighs about 2 kg, and its cheek band is a thin line of white feathers starting from the eye and running under the chin to the opposing eye. The body band is a relatively thin band running closely along the main body of the black back, from the base of the leg upward across the chest and to the other leg. The Humboldt penguin is larger at 65 cm and 4 kg, and the banding is wider both in the face and in the body. Both have patterned black feet. The Magellanic has speckled black and pink feet. The body band is broad, as is the face ring, which is so broad that it results in a double band across the chest. This second band connects from the lower part of the neck across the chest. It is present also in the Galápagos, but in this species the facial ring is so thin that the band is not obvious. The Magellanic penguin is about 70 cm in length and weighs 3.5 kg. In addition to these South American birds, there is one representative in Africa. The African or black-footed penguin has black feet and a broad body band that extends across the chest with a striking gap. It is 70 cm in length and weighs 3 kg. All of the species in this group have a bare patch of skin somewhere over the eye, except that in the Humboldt, much of the bare spot is in front of the eye and extends to below the mandible.

Northern auks, aquatic divers of the Northern Hemisphere that have many traits of penguins but are not related to them, including the use of their wings for underwater propulsion. *From left to right:* dovekie, Atlantic puffin, and thick-billed murre.

Why are there no penguins in the Northern Hemisphere?

The short answer is that the waters are too warm in the tropics and that sea lions and fur seals (Otariidae) species were expanding their range of distribution at about the same time; they out-competed penguins for food and space and preyed upon them. Land predators prevented mainland nesting, but that is true for penguins breeding on the mainland of South America and South Africa, also. That leads us to an even shorter answer: We do not know. However, penguins evolved in the tropical and subtropical waters. Here were some of the largest penguins ever, and why they did not go north is not clear. What we do know is that the present-day penguins are all a product of cold-water origins and species distribution and an increase in the number of species. By this time small cetaceans had already spread out throughout the world's oceans, and, concurrently with the penguins and the pinnipeds (sea lions and seals), were becoming more diverse as they moved into new habitats occurring with the changing conditions of the Cenozoic. The giant, warm-water penguins of the Eocene and Oligocene had probably disappeared. Why the present-day penguins never penetrated beyond the Southern Hemisphere is a mystery, but the fact that they were evolving in cold waters and that there were large marine mammal competitors already in the tropics and the Northern Hemisphere did not help.

Macaroni penguin colony. Note the concentration and uniform distance between nests.

Where do penguins live?

Penguins range throughout the oceans south of the equator, with some qualifications. They are most likely to be seen near islands or the continental coasts in cold waters. Offshore and off the continental shelves, they become scarce. The cold Pacific, Atlantic, and Southern Oceans are the most likely regions to see penguins, and penguins are not likely to be seen at all in the warm waters of the Indian Ocean.

The greatest concentrations and diversity of penguins is in the cold-temperate to sub-Antarctic islands. The most abundant species, which forms the largest colonies, is the macaroni penguin, found across the entire eastern sector of the Southern Ocean from Tierra del Fuego and the Antarctic Peninsula to Heard Island, or 90°W to 90°E. The sub-Antarctic is also the region of greatest diversity, with penguins occurring on almost every island, from those at the tip of South America to the constellation of islands around New Zealand; 10 of the 17 species live in that area. There is one significant gap, and that is the "No Man's Ocean," a part of the Southern and South Pacific Oceans from just west of Cape Horn to just east of New Zealand, where there are no islands. Toward the equator penguin distribution declines abruptly, with no colonies and probably no wanderers in the Indian Ocean off the east coast of Africa or the west coast of Australia north of Perth. The range north is more extensive in the south Atlantic and the south Pacific, where the African penguin has colonies as far north as

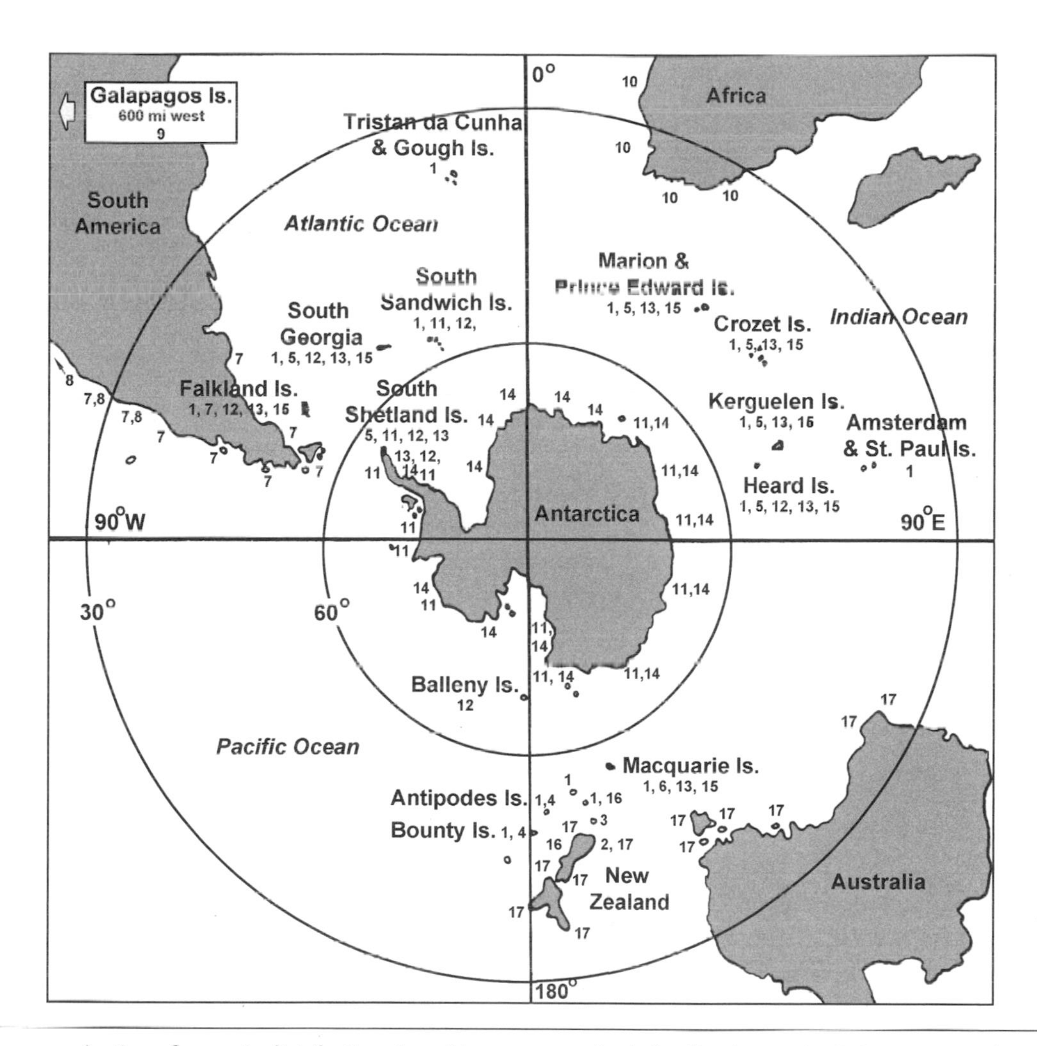

Polar map projection of penguin distribution. 1, rockhopper penguin; 2, fiordland penguin; 3, Snares penguin; 4, erect crested penguin; 5, macaroni penguin; 6, royal penguin; 7, Magellanic penguin; 8, Humboldt penguin; 9, Galápagos penguin; 10, African penguin; 11, Adélie penguin; 12, chinstrap penguin; 13, gentoo penguin; 14, emperor penguin; 15, king penguin; 16, yellow-eyed penguin; 17, little penguin.

about 25°S on the west coast. The eastern Atlantic seaboard of Argentina has Magellanic penguin colonies to about 40°S, but this species reaches to 30°S on the west coast of South America and overlaps slightly with the Humboldt penguin, which extends nearly to the equator along the coast of Peru. Then there is the Galápagos penguin, which exists in isolation along the coasts of mostly Fernandina and Isabella Islands of the Galápagos Islands, on or near the equator. It is the only penguin that has made it north of the equator, but its range barely reaches this far north location. On the western side of the Pacific, there are no penguins along the eastern coast of Australia. The little penguin has an extensive range from Perth in western

Australia, along the entire southern coast, and beyond the Tasman Sea to the entire coastline of New Zealand.

Clues to the range limitations come from the banded penguins, which are the most temperate and northerly occurring of the groups of penguins. The northern range extension occurs in waters with cold currents sweeping along continental shores. This causes upwelling that brings nutrients to near the surface and draws large schools of small fish, most notably the anchoveta along the Chilean and Peruvian coasts and sardines along the African coast. In addition to this important food resource for the Humboldt, Magellanic, and African penguins, the cold water also results in frequent cloud cover along the shoreline that keeps the coastal temperatures moderate. Soil conditions are suitable for burrowed nesting sites or small caves, both of which provide shelter from the otherwise intense heat that occurs during the day when the overcast burns off. This is especially critical when adults are incubating eggs or brooding chicks. For these species there are few suitable islands where the nesting habitat is away from terrestrial predators. The Galápagos penguins are exceptional in this regard. They live where there are no land predators, there are many caves along the shoreline that are suitable for nesting, and there are schools of fish near the shore. Even in such an isolated habitat, the availability of suitable nesting sites and the unreliable food source limit abundance. There are also the extreme oscillations in weather when El Niño occurs cyclically about every seven years. During the most severe El Niños, the population of penguins is depleted. In the 1982–83 El Niño, only 23% of the population estimated in 1980 survived. Up to 2003 it has not returned to pre-1983 levels.

All of those factors play a role in where penguins live, but probably the most significant features farther north are the high temperatures of surface water and land along the shoreline, coastal areas that have little overcast, numerous and varied land predators, and the width of the tropical barrier in both the Pacific and Atlantic Oceans. For example, there is no evidence of penguins ever being on the tropical islands of Mauritius or Réunion before or after humans arrived and eliminated much of the island fauna, including the dodo. This is also true of isolated subtropical islands in the Atlantic and Pacific Oceans such as Helena, Bounty, and Easter Islands and the eastern coast of Australia along with its scattered groups of islands extending from New South Wales to Queensland. All of these islands lacked land predators and supported other types of seabirds' colonies, and some may even have caves that would protect penguins on nests or brooding chicks.

Are there fossil penguins?

There are more known fossil penguin species than extant species. If the Eocene period had not become known more famously and earlier as the age of whales and land mammals, it could have been considered the age of giant penguins.

When did penguins first evolve?

Based on molecular dating, penguins are as old as 71 million years, and they may be as old as 90 to 100 million years, but so far there is no fossil evidence for this. During the late Cretaceous in the region of Gondwanaland, as that supercontinent was breaking up, and with the fragmentation into the Southern Hemisphere continents, the penguins did diverge and spread out. Some of the diversification was probably a result of loss of predatory competitors as the ichthyosaurs and plesiosaurs, large marine reptiles, disappeared at the end of the Cretaceous. This was also during the time when the large marine predatory birds Ichthyornis and Hesperornis became extinct; these may have been in competition with penguins and suppressing their geographic expansion.

What is the oldest fossil penguin?

Al Mannering found the oldest known fossil penguin in the Greensand formation, near the Waipara River, South Island, New Zealand. Based on the location, it is estimated to be early Paleocene and approximately 60 million years old. It is part of a generic group of two species and is the largest of the two; it was nearly as large as the present-day emperor penguin.

What is the largest fossil penguin?

There are two extinct species of penguins that were near the same height of about 165 to 170 cm. They are estimated to have weighed about 135 kg, or about five times the body mass of the emperor penguin, the largest present-day species. One of these two large fossil species was found on Seymour Island, Antarctic Peninsula, and the other was found on South Island, New Zealand. It should be noted that the size of these and most other fossils was estimated from a few bones, in particular the humerus or the tarsometatarsus. These bones can vary between different species of penguins, so the estimate of size of most fossil penguins is coarse.

Chapter 2

Form and Function

What are the largest and smallest living penguins?

When most people think of birds, they usually think of the canary-in-the-cage kind that are used to provide an early warning for all kinds of problems. The idea is that the birds will succumb before the humans. In reality, many birds are large and robust, and some are able to cause serious injury to a person with a single blow. The largest of the penguins are in this category, the ones that belong to the genus *Aptenodytes.* They are capable of breaking your nose with a single blow of their wing or ripping your pants off with the heavy claws of their feet if you hold them in a bear hug. All penguins have a seasonally variable body mass, and the emperor penguin male has a body mass of about 40 kg shortly before breeding and entering the winter fast or just before the summer molt. At the end of the molt or after the winter fast, their mass may be as low as 16 kg or 20 kg, respectively. The common "athletic" condition when penguins are foraging for food and feeding their chicks over the five-month winter-through-spring period is about 25 kg for both the male and the female. The closely related king penguin, the second-largest penguin, has half the body mass of the emperor penguin. The smallest penguin is the little blue penguin, which at 1 kg is about one-thirtieth of the weight of the emperor penguin. Penguins all go through substantial changes in body mass depending on the season and their activity.

Emperor penguin trio walking on ice.

What is the basic structure of penguins?

Across the wide array of bird diversity, the external variability of shape, size, and color is a natural wonder. Birds range in size from the 150 kg ostrich to the 2 gram bee hummingbird. The length of the Australian pelican at 49 cm is the longest of any bird. In terms of proportion, the sword-bill hummingbird's bill is about two-thirds the length of its body. The colors and shapes of feathers range from the uniform black of crows, with not a contour feather out of place, to the male birds of paradise with their fantastic colors and feather shapes prominently displayed on their heads, tails, and breasts.

Diving birds are much more conservative in their body form, which is driven by the physical nature of the medium in which they make a living; hydrodynamic forces illustrate how form follows function. All aquatic birds' bodies are spindle-shaped, and the penguin has the most perfect spindle of them all. Feathers are often black and white in aquatic birds, and they always follow this color theme in penguins. Feathers are also dense, and those of penguins are the densest. Penguin bills are often short and conical, with varying degrees of curvature. The long neck, containing 15 cervical vertebrae in the emperor penguin, is coiled while swimming to maintain the spindle shape but can extend rapidly to strike at prey. Dominating the torso are the large pectoral muscles, the main engine of propulsion. Internal organs are tucked mainly into the protective shell of the rib or thoracic

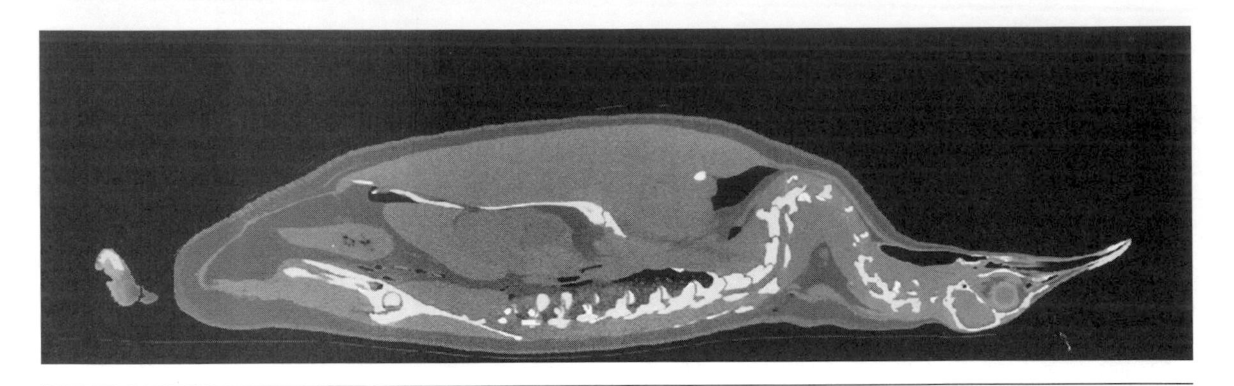

This gray-and-white profile section shows the thin skin, the thicker fat layer, and the large area the breast occupies. Also note the area of the eye relative to the brain. The computer tomographic scans were made from a 27 kg adult male within 24 hours of death. Courtesy Miriam Scadeng, University of California–San Diego.

cage; part of the stomach, the small intestine, and the entire pancreas are at the bottom of the abdomen.

The wings are the best external landmarks for the internal anatomy of the penguin. The wings insert into the coracoid bones at the top of the rib cage. The tip of the wing reaches to the knees, just above the base of the thoracic cavity (between the neck and the abdomen). The lungs extend from near the insertion point of the wings to midway down the wing length. Cervical air sacs occupy the neck from the middle of the cervicals to the clavicle bones at the top of the thoracic cage. The front and back air sacs extend from the upper edge of the thorax to slightly beyond the lower edge. Paired abdominal sacs begin at the base of the lung and reach to the bottom of the abdomen and the vertebral column.

What is the metabolism of a penguin?

The metabolism of a penguin ranges widely depending upon the activity, stage of development, and size of the species. For comparison, an adult emperor penguin has a resting metabolic rate that is about the equivalent of a 50-watt incandescent light bulb. The resting metabolic rate of the little blue penguin is about equivalent to a 15-watt bulb. However, wild animals, especially aquatic birds, are seldom resting. When traveling to forage, their metabolic rate is 2 to 4 times as high as that of a resting bird. The maximum rate for an emperor penguin, the only penguin for which it has been measured while swimming, is nine times that of a resting bird. This is not great compared with how a thoroughbred racing horse's metabolism varies; the horse may reach 40 times its resting level during a sprint. Most aquatic birds and mammals do not have a large metabolic range.

Emperor penguin trio tobogganing on snow. The wings are used for balance unless at speed; then they strike the snow in unison to aid the foot propulsion.

In fact, penguins are not animals of great energetic output; they are characterized by energy conservation instead. Penguins have to sequester and use their oxygen stores judiciously while diving to depths and during long fasts. The most extreme in long-term energy conservation is the emperor penguin. This topic is discussed more in chapter 5, but consider that incubating males must endure a total fast of about 120 days in winter. Yet, under these conditions the males reduce their fasting or incubating metabolic rate to about 25% lower than the resting rate. They do this by frequently huddling together. The group behavior of huddling, along with limited activity in or out of the huddle, results in an energy saving essential for them to sustain the fast throughout the incubation period without running out of internal energy stores.

Do penguins have teeth?

Like all other present-day birds, penguins do not have teeth. The same is true of all known extinct penguins as well. As far as we know, the last birds to have teeth lived in the Cretaceous period, some 90 to 100 million years ago. However, penguins' tongues have spiny protuberances that point backward toward the esophagus. These function not for chewing but to prevent the prey, such as fish, from resisting being swallowed. Most prey are swallowed whole, unless they are so large that they must be torn apart. In most cases the fish are manipulated so that they are swallowed headfirst. The penguin's lower mandible has the capacity to expand to a large width, and this also helps in swallowing large prey.

Do penguins sleep?

All penguins, while they are on land, appear to sleep, and sometimes they sleep very soundly. I have been able to approach close enough to sleeping Adélie and emperor penguins to touch them before they react to my presence. Recently electroencephalogram measurements of birds have shown two major stages of sleep, which are shared with mammals. The longest, the non-rapid eye movement stage, is the deep, slow-wave phase, and rapid eye movement (REM) sleep is when most dreaming occurs. It represents about 20% of the total sleep. In some mammals, such as whales, there is also unihemispherical sleep, during which half of the brain remains awake. This must be an important adaptation not only for alertness to the approach of predators in an environment where there is no place to hide, but for the constant need to breathe at the surface of the water. Similarly, there are some species of birds that land infrequently and some that remain on the wing for months at a time. Swifts are one example of a long-duration airborne bird. Although unihemispherical sleep is unknown in most birds, it seems plausible that some birds would experience it. When penguins are at sea for long periods of time, they might and should have this capacity under appropriate conditions. Being buoyant, they can float at the surface, but at times they must still be alert for predators from above or below. Also, in rough seas they must be alert to waves washing over them and wake up before attempting to breathe while under water.

Can penguins see color?

The eye of the penguin is adapted primarily for nighttime or low light levels. It is made up of two types of light-sensitive cells, or photoreceptors. The rods, which provide mainly black-and-white vision, are the most light-sensitive elements. The cones perceive color and function mainly in daylight. Usually in vertebrates the cones are concentrated in the central part of the visual field and the rods in the periphery. That is why it is better to look at faint stars from the periphery of the eye. In the Magellanic penguin, the rods are distributed uniformly over the entire retina, but the cones decrease toward the periphery.

There is little published information on penguin vision, even though penguins' amphibious life style suggests some unusual adaptations. Penguins are able to see well above and below the surface of the water and in low-light conditions, and they are visual predators. The low-light sensitivity is especially important for the deep divers, such as the king and emperor penguins, which dive to considerable depths. In that black and gray world, the level of light can decline to the equivalent of starlight, and the birds

Sleeping Adélie penguin.

must make use of biological light emitted by their prey. This reliance on biological light seems a logical capacity to have, since about 90% of animals that live at depth have some means of producing light. Most of these organism produce in the blue-green spectrum.

Even on the surface, there is little color in the oceanic environment. It is a different matter when penguins are ashore. Not only is there color in the habitat for them to sense, but most penguins have color in their feathers and eyes. These colors help a species determine the gender and the age of individual penguins and even the condition of a penguin's immune system. For example, king penguin adults have strikingly colored auricular patches, which change from no yellow in the juvenile to the bright yellow of the adult. The bill also is black above and the mandibular plate (lower jaw) is orange to maroon on the proximal half of the beak; it may change its shade depending on whether a penguin is in water or in air. This lower orange bar of both *Aptenodytes* species also has ultraviolet reflectance properties in the adults. Both the color and the ultraviolet reflectance are important in the pairing of adults, as is the yellow auricular patch. In king penguins the yellow patch also indicates to other king penguins an individual's body condition and whether the penguin is an early seasonal breeder. Early breeders are usually the most successful, and this feature is to some degree indicated by the brightness of the yellow patch. The size of the yellow patch is most important for its perception by king penguin females, and the ultraviolet reflectance is important to both sexes. Thus, in this species there is some difference in mate choice between the male and the female. There is also evidence that these ornamental traits are an indicator of the aggressiveness of the mate, and they may result in better location of the nest site. In short, the ability to see color, among other traits, promotes the ability for sexual selection in king penguins, and likely in other species of penguins as well.

Do penguins molt?

The molt occurs once a year, soon after the chicks fledge. If an immature or adult bird has not bred, then molt in breeding adult may occur sooner than it does for those producing chicks. It is imperative that food be abundant where penguins forage shortly before the molt, so that they may increase their fat stores substantially. Emperor penguins increase their body mass during this period by about 30% to 50%, compared with their body mass when they are nurturing chicks. This may be extreme because the emperor penguin has the longest molt period of any penguins and because emperor penguins rest on sea ice for 35 to 40 days during the process. They also may travel a considerable distance from the time after leaving the chick to settling on a floe or fast ice to molt; fast ice is attached to the shoreline and does not move. The distance of travel to the molt area is often more than 1,000 km in a straight-line distance, but penguins never travel in a straight line during foraging trips, so the overall distance is easily two to three times the "as a crow flies" metric, and that does not include the up-and-down distance under water in search of prey.

Few emperor penguins molt at the colony site, because they reproduce on annual sea ice and it usually disintegrates shortly after the chicks fledge. For some reason, this is also generally true for Adélie penguins, even though they breed on land. During the month of travel to the molt area, both emperor and Adélie penguins must also be feeding enough to gain the increased body mass necessary for the monthlong fast. A somewhat modified pattern is followed by Adélie penguins, which depart their colonies in mid-February, during or after the time when the chicks are fledging. They also seek out stable pack ice, where the floes will last until their 20-day molt is finished. However, because they breed on land, there is not the uncompromising constraint on leaving the breeding colony, and some will molt there. Most other penguins, especially sub-Antarctic and cold-temperate species, will depart the colony for an extended foraging period that may last from about two weeks to a month before returning to the colony or a nearby area to molt. By the time penguins arrive to molt ashore, the new feathers have already begun to develop, and over the next period of time all old feathers will be shed before the birds return to sea with an entirely new coat of feathers. For a detailed description of molt in terrestrial birds, see the website All about Birds at www.birds.cornell.edu/AllAboutBirds/studying/feathers/molting/document_view.

Rockhopper penguin molting.

Can penguins run?

Some species of penguins can run. The most extreme nonrunner is the emperor penguin. Since the legs of emperor penguins are covered by skin to nearly the tops of their feet, they cannot run. It would be like a person trying to run with his pants down. Other species can run, after a fashion. None have a long stride, and they may not have both feet off the ground at the same time, so it would be defined as a fast walk; but even a fast walk is impossible for emperor penguins. Many species come ashore on beaches and seldom if ever leave those surface conditions, since they nest on nearby and elevated nearshore habitats. Those penguins may walk at a leisurely pace to the colony, or if motivated they may increase their rate and "run." Examples of these species are the Magellanic, gentoo, and king penguin. Others land on steep rocky faces and must ascend cliffs. In this case hopping is an efficient way to move; the best example of such penguins is the rockhopper penguin.

Walking Adélie penguin.

Why do penguins have flippers?

The wings of penguins are their primary mode of propulsion. But instead of flying in air, penguins use their wings, which are reconfigured from large flapping wings into small rigid flippers (see chapter 1), to fly under water. In addition to being penguins' propellers, the wings provide stability for gliding and serve as ailerons for directional control. For propulsion they function as a foil, as an aircraft propeller, not a paddle. With each up-and-down stroke of the wing, it is twisted so that the main vector of lift is projected forward. This enables penguins to travel rapidly through the water, although at a much slower speed than a flying bird, because water is much denser than air. However, because the body of a penguin has close to neutral buoyancy, almost all of the lift from the wings can be directed to forward travel. While swimming, penguins can travel at a relatively low cost of energy compared with flying birds, which must overcome the force of gravity.

Swimming Galápagos penguin.

Can penguins breathe under water?

Neither penguins nor any other bird or mammal can breathe under water. The metabolic rates of both birds and mammals are so high that they would need a very extensive skin surface in order for enough oxygen to diffuse across the membrane and enter the blood. A few reptiles may have a small amount of respiration while under water, such as the yellow-bellied sea snake. This can help extend the snake's ability to hold its breath, but it does not enable it to rely solely on breathing through the skin.

Do penguins need to drink?

Most of the year, Antarctic penguins are at sea and rest on pack ice or icebergs where some freshwater might be available to them in the form of snow. Sub-Antarctic penguins all breed on islands where there is much snowfall or rain, and freshwater is also readily available in streams and ponds. Cold-temperate penguins of South America, South Africa, and Australia come ashore to rest and breed in desert conditions. These are shorelines where cold currents such as the Humboldt and the Benguela produce little or no rain; where the Chilean Humboldt penguin lives, the Andes Mountains produce such an effective barrier and rain shadow for storms from the east that the Atacama Desert is the driest place on earth. In these cases, no reliable freshwater is available.

However, all penguins can probably drink seawater, because of the salt gland that is embedded in the supraorbital bone of the skull (the bone above the eye socket). This gland is present in all seabirds and some land birds. In seabirds it can extract any excess salt from the blood and concentrate it in a

Chinstrap penguin eating snow.

solution that is many times saltier than the salt in the bird's blood. This end product, like tears, is excreted from the gland, flows down a groove in the bill, and drips off the tip of the bill. Because of this capacity to produce a concentrated solution extracted from the blood, without reliance on the kidneys, they have no or little need for freshwater. Most seabirds, including penguins, meet their water needs from their food of fish, krill, and squid, all of which are at least 75% water. On warm days, or when fasting, penguins will drink freshwater if it is available. In addition, the polar penguins are commonly seen eating snow or even chunks of ice, especially on warm days. The reason may not be to slake their thirst but to lower their core temperature.

Chapter 3

Penguin Colors

Why are penguins black and white?

The standard reason given for the striking black-and-white coloration of penguins is "camouflage." The wide prevalence of this color pattern, a black back and a snowy white chest with a sharp lateral boundary between the two colors, suggests strong selective pressure for certain gene combinations. All penguins have this general pattern, and only the banded penguins (Humboldt, Galápagos, Magellanic, and African) deviate slightly. In these four species there is a double lateral boundary in which a thin lateral, black-to-white stripe runs the length of the body from below the beak to the top of the legs. The band of the Galápagos penguin is the thinnest, and that of the Magellanic exhibits two neck bands rather than one. The main band starts just above each leg, and a large U shape forms as the two parts of the band meet under the bill. The extra band of the Magellanic extends from the back and wraps around the neck above the body band.

The black-and-white color pattern, like that of many oceanic vertebrates, is believed to play a role in countershading. While the bird is at sea, this affords a cryptic silhouette from below and from above. From below, it is difficult to see the bird because its white belly blends with the bright background of the sky above the sea. From above, the black back makes the bird nearly impossible to see against the bottom or the dark of the depths below. These features confer an advantage for the birds when they approach prey from below and also disguise their presence from predators searching from above. The reason for the extra lateral stripe in the *Spheniscus* group is unknown. One proposed suggestion is that the exaggerated color pattern of banded penguins (*Spheniscus*) is so conspicuous that it tends

Adult Magellanic penguin showing the distinct lateral body stripe.

to break up the schools of the fish they pursue. In either case, the birds that vary from their species's color pattern, the ones that are all white, or all black, or somewhere in between, are at an important disadvantage compared with their normally colored companions. Curiously, an ancient and large penguin of Peru that existed about 36 ma (million years ago) was not black and white. A very technical assessment of its feather structure and the presence of light-absorbing organelles called melanosomes, using scanning electron micrography, indicate that it was gray and reddish-brown.

What causes the different color patterns of penguins?

All color characteristics of penguins are a result of genetics, most likely along with strong selection pressure. The selection process may be for discrimination of age and sex. For the banded penguins (*Spheniscus*), which have no color except black and white, with the exception of the bare, pink skin around the bill in the adults, the only variation is a small amount of black spotting on the chest and belly. However, the juveniles of this group have no banding, and they are easily distinguished from the adults. The juveniles of other penguin species are discernible from the adults by a modification in the extent and intensity of the black back, which either does not extend under the chin or is gray instead of black.

One species in which the adult and juvenile look alike is the little penguin. The feather color and pattern is the same, and even the eye color of the juvenile matches that of the adult. The chinstrap penguin also shows no difference between the adult and the juvenile; both have a simple black stripe under the bill, and the eye color is the same.

The royal penguin, which has one of the more modest crests of the crested penguins.

In most other species there are differences, such as additional shades of yellow for feathers on the head and different iris and bill colors. Crested penguins (*Eudyptes*) are the most elaborately feathered penguins. Their black-and-white countershading is conventional, but the crests are not, and the crests are distinctive for each species. The rockhopper has a yellow eyebrow (superciliary stripe) that turns into a spray of feathers at the back of the head (superciliary crest plume). The erect crested penguin's eyebrow is in the form of a plume like that of a rocker's hairdo, or feather-do. However, when the crest is not erect, it is a sleek chevron-shaped racing stripe. The eyebrow feathers of fiordland penguins, when laid back, also look like parallel racing stripes. When erect, the tail end of the stripe forms a spray. This penguin also has a small striped pattern of white on the cheeks. The Snares penguin's crest is similar in shape, with more of a spray when erect. The sister groups of macaroni and royal penguins have similar long eyebrow feathers that form an untidy spray, hence the name of macaroni. Again, this group has distinct color patterns, which the immature birds do not possess. The elaboration of the superciliary stripe and the crest with maturity probably facilitates the expression of emotions related to sexual displays. Finally, the *Aptenodytes*, which includes the king and emperor penguins, has bright yellow auricular (ear) patches. Although there is graded decline in the intensity of the yellow on the breast for both species, the emperor penguin is the only penguin adult with a soft golden breast that extends nearly the length of the chest and belly.

The function of these varying colors is unknown, but they likely play a role in age identification. Little is reported on most penguins, but the one species in which much work is reported is the king penguin. Studies show that the yellow color of the auricular patches and the ultraviolet properties of the bill relate to mate selection. Some of these colored features also provide information on body condition and age of the prospective mate, which are especially important factors in partner selection.

What color are penguin feet?

The *Aptenodytes* penguins, the two largest species, have black feet. The banded penguins characteristically have black or mottled pink-and-black feet. All other penguins have pink feet. Gentoo penguins' feet are more orange than pink. In the pink-footed birds, the feet become pale if the penguins stay ashore for an extended time.

What color are a penguin's eyes?

Guess what? Adult yellow-eyed penguins have bright yellow eyes. Somewhat similar are the adult little or white-flippered penguins, with pale yellow to gray eyes. All adults of the *Eudyptes* group have dark red eyes. Rockhopper adults have the brightest red eyes of any penguin species. Both gentoo and Adélie penguins have dark brown eyes, while the chinstrap penguin's eyes are light brown. King and emperor penguins have dark brown to black eyes.

Do penguins' color patterns change as they grow?

From year to year color patterns change in the juvenile until it is about three years old. The most striking variations in color are of the bill, the eyes, and the feather patterns of the head and neck. The greatest change occurs after the juvenile molt. Both king and emperor penguins have a faint yellow to soft pink stripe along the mandibular (lower) portion of the bill, which will vary in depth of color depending on how recently the bird was in the water. The mandibular stripe is not present in the all-black bill of chicks and juveniles. Over the course of the first year, it develops into a muted pink or yellow in the juvenile bird and deepens in color as the bird matures to adulthood at about age three. The bright yellow auricular pattern of these two species is not present in the recently fledged juvenile. It becomes more apparent after the second- and third-year molt. Similarly, the black chin area of the adults is less developed in the king, and in many emperor penguins it is not present at all until the second year. However, there is variation in this head patterning.

Brush-tailed penguin juveniles show a different pattern within each species regarding the head pattern. The chin stripe of the chinstrap penguin is present in the juvenile, whereas the black chin of the adult Adélie penguin is absent in the juvenile. The only variation from the juvenile to adult gentoo is that the supraorbital (above the eye socket) white eye patch increases in size from the juvenile to the adult.

In the yellow-eyed penguin, the juvenile eye is a fainter yellow than the

The black feet of the king penguin.

Juvenile Adélie penguin. Note the white chin and the lack of a white eye ring.

adult eye. In addition, the dark chin is faint or absent in the juvenile, and there is no yellow crown or eye stripe, which is so apparent in the adult.

The most varied head decorations of all penguins are in the adult crested penguins. No such patterns are present in the first year of the immature penguins. In the juveniles the bill is not the robust red of the adult but is black to faintly red. Exceptions are the macaroni and royal penguins, in which the juveniles have a robust red bill; in the macaroni there is a faint amount of spray in the crest on the back of the head. In all others there is an eyebrow stripe, which is greatest in the fiordland juvenile and least in the rockhopper.

In none of the banded penguins is the band present in the first year. Also absent is the featherless area (apteria) at the base of the bill. This pink area is well developed in the adult from the bill base to the eye. It is largest in the Humboldt penguin.

Do a penguin's colors change from season to season?

Unlike many birds, whose colors change dramatically during the breeding season (for example, male ducks), the color of the penguin is fixed throughout the year until the annual molt. The color pattern for adults remains the same year after year, but the young birds gradually phase into the adult plumage after two to three years. The exception to this rule is the little penguin, in which the juvenile is similar to the adult. Indeed, except for subtle differences in size, the male and female penguins of all species are indistinguishable by sight for humans. Color and pattern depend on the eye of the beholder, and penguins see color and pattern quite differently from the way humans do (see chapter 4).

Is there much geographic color variation in a single penguin species?

There is most likely some color variation in widely distributed penguin species, but surprisingly little is known about the subject. One of my major sources for reference, *The Penguins*, by Tony Williams, makes no mention of color pattern variation. In fact, little is said about penguin colors except in the color drawings for each species. Those drawings are excellent, but they do not deal with geographic variation. Frank Todd's field guide *Birds and Mammal of the Antarctic, Subantarctic and Falkland Islands*, shows six variants of the little penguin, if the little blue penguin is lumped with this group. The differences are subtle except in the white-flippered penguin, which has a distinct white border along the leading and trailing edge of

Moseley's rockhopper penguin with the exceptionally long crest feathers.

Instead of white spotting, as described in the text for the emperor penguin, this yellow-eyed penguin has patches of solid white on the back of the wings and on the back.

its wings. The white-flippered penguin is considered a separate species by some taxonomists.

Of the eight subspecies of the circumglobal rockhopper penguin, the illustrations in Todd's book show some subtle differences in the eyebrow plume in three subspecies. The crest of the short-crested rockhopper penguin is the most modest of the three crested penguins. It occurs on the Falkland Islands and the Cape Horn region. The crest of the eastern rockhopper penguin is intermediate in length, compared with that of the long-crested rockhopper penguin, which is gaudily different from the others with an exceptionally long and thick crest. However, few avian biologists or penguin enthusiasts are likely to see this subspecies, since it occurs only on the South Atlantic and Indian Ocean islands of Tristan de Cunha, Gough, Amsterdam, and St. Paul. These are some of the most remote islands in the world. A person is more likely to see the eastern rockhopper penguin, with its widespread distribution on all the sub-Antarctic Islands of the South Pacific and southern Indian Oceans.

Finally, Todd differentiates the northern gentoo penguin from the southern gentoo penguin by the larger and more orange bill of the northern subspecies, as well as its longer flippers and feet. The northern occurs on all of the sub-Antarctic islands, while the southern occurs on the Antarctic Peninsula and the nearby Antarctic island groups of the South Shetland, Orkney, and Sandwich Islands.

Other types of color variation, which may not be related to geographic distribution, are color morphs (variants) that range from leucistic (reduced pigment) to melanistic (increased pigment) birds, with a host of variation in between (all white or black to abnormal patterns of white and black). I spotted my first all-black king penguin in Saint Andrews Bay, South Georgia, in 1985, and since then I have seen a few on Isle de Possession of the Crozet Archipelago. Others have reported sightings at a few colonies in different island groups. At least one all-black Adélie penguin was observed, but the location was not reported. Those are the only all-black penguins known to me, but likely there are others in other locations and species. My son Carsten and I observed our first and only all-white emperor penguin at the Cape Washington colony, Antarctica, in 1996. Before and after that sighting we have seen several white color morphs with different degrees of white spotting on the black back. The only all-white king penguin was observed and photographed by Frank Todd at Saint Andrews Bay, South Georgia. Of the smaller penguins, there have been several sightings of all-white or nearly all-white brush-tailed penguins. Much of the likelihood of seeing color variations is a matter of sample size. The larger the colony being observed, the more likely that some form of color variation will occur.

Chapter 4

Penguin Behavior

Are penguins social?

Like any animal species that reproduces sexually, penguins exhibit some degree of social behavior. The most striking behavior displays of penguins are all about breeding, and they can be categorized as (1) agonistic, or aggressive, (2) sexual, (3) vocal, (4) incubation, and (5) crèching (the young forming a group). Much detail is available about the social activity of some penguin species while they are ashore, because they are synchronous in their behavior (that is, they are ashore at the same time) and they congregate in large breeding colonies, totally exposed and consequently highly visible. They are also not shy and can be observed at close range. The secretive nesting species that burrow or nest in dense vegetation are often not found in large colonies.

All penguins except king and emperor penguins build some sort of formal nest. The nests are usually above ground, but some species breed on surfaces that are suitable for burrowing. Burrowers all belong to the genus *Spheniscus*, but in some cases a colony will include both burrowers and surface nesters. Some *Spheniscus* will also take advantage of caves and crevices. The Galápagos penguin rarely if ever nests in the open. Those species that nest on the surface are more subject to aerial predation, and the birds on the periphery of the colony are the most vulnerable of all. An additional advantage of burrows, caves, and crevices is protection from solar radiation. This is especially true for the Galápagos penguin, which lives near the equator, and the northern colonies of the African, Humboldt, and Magellanic penguins. With the exception of the macaroni and the royal penguin, the *Eudyptes* genus all nest in small colonies in caves, crevices, or heavy vegetation.

The little penguin, because of its size and because it nests on large islands such as New Zealand and on the continent of Australia, is obligated to nest in burrows, caves, and crevices for predator avoidance; it also usually travels to and from the colony during the dark. The extremes are emperor penguins, which must congregate and huddle during the winter blizzards when they breed, and the yellow-eyed penguin, which prefers to have much space around its nests. It has been said that the yellow-eyed penguin will not nest within sight of another nesting pair. However, it is unclear whether this is a preference for solitude or whether these birds simply prefer heavily vegetated areas and cannot see one another as a result, even if their nests are close to one another. Penguin behavior ranges from violent aggression while breeding, in at least the king, chinstrap, Adélie, and macaroni penguins, to the extreme cooperation of emperor penguins during the dark winter months. Despite the climate where emperor penguin colonies occur, it is a Camelot, "where never a harsh word is heard." All penguins use both vocal and visual displays to attract mates, to indicate recognition of each other, and to intimidate competitors. They also use the calls and visual displays to assess partner fitness.

While at sea, penguins' sociality is less clear. They are more likely to be seen in large groups than solo. Also, if observations are made near a colony, groups are more likely to be seen as the penguins come together during arrival or departure, often in large groups, which may disperse farther offshore. After they separate from the colony to forage for themselves and the chicks, and during the long goodbye before and after the molt, when they depart the colony or molting area for several weeks to a few months, their sociality is less certain. I have seen both groups and solo animals of king, emperor, and Adélie penguins at sea and distant from any islands or colonies. Most remarkable was a single Adélie penguin who was seen sleeping on a floe in the Ross Sea at 74.3°S and the 180° meridian, on May 17, 1998. This was well south of the sun horizon; there was 24-hour darkness or civil twilight at that time. Four other solo Adélie penguins were seen on the same cruise, three of them also in the twilight zone. Most other sightings were of birds on the pack ice at night and in groups ranging from two to 190 birds. Many of these were in the twilight zone. Similarly, there were several sightings of emperor penguins. In addition to several solitary birds, there were groups ranging from two to 35 birds. Most were seen in the twilight zone. Notable among these sightings were three juveniles at 69°S on May 10. From our observations I conclude that Adélie penguins are mostly "flocking" birds and that emperor penguins tend to be small-group or solo birds. In contrast to these winter observations, the fledging (leaving

Emperor penguins underwater. Note the wing position at rest, the streamlined body contour, and the bubble trail streaming from the head and back of the ascending bird. Photo by Paul Ponganis.

Bill profiles. *Clockwise from upper left:* emperor, macaroni, gentoo, and Adélie penguins. Note the nares opening at the base of the bill where the feathers begin. Also note the groove in the bill where the fluid from the salt gland flows down to the tip of the bill.

Comparative penguin eye colors. *Clockwise from upper left:* king, rockhopper, yellow-eyed, and little penguins.

Penguin feet. *Clockwise from upper left:* emperor, Galápagos, Adélie, and gentoo penguins.

All crested penguin heads. *Clockwise from upper left:* macaroni, royal, rockhopper, erect crested, fiordland, and Snares penguins.

Heads and shoulders of banded penguins. *Clockwise from upper left:* Galápagos, Humboldt, Magellanic, and African penguins.

The golden chest of the emperor penguin, the only species with such a cast to its chest color.

An Isabella color variation of an Adélie penguin.

Yellow-eyed penguin at its nest.

Magellanic penguins coming ashore.

Heads and beaks. *Clockwise from upper left:* Gentoo, chinstrap, Adélie, Magellanic, and Humboldt penguins. The Magellanic and the Humboldt penguins are mainly fish eaters, the chinstrap and the Adélie penguins are primarily krill predators, and the gentoo penguin is perhaps the most versatile, with a range from krill to fish. Here also is a good comparison of the feathering on the face, as the heat-tolerant birds have fewer feathers on their faces than the cold-resistant polar birds do.

Rogues' gallery of penguin predators. *Clockwise from upper left:* giant petrel, south polar skua, brown skua, male Antarctic fur seal, and leopard seal.

Pair of rockhopper penguins.

King penguin with egg. Like the emperor penguin, the king penguin supports its egg on its feet, but the king penguin defines a nesting territory and defends it vigorously from neighbors.

This small group of nesting king penguins shows the even spacing they maintain, even though there is no formal nest such as the chinstrap penguin builds.

Magellanic penguin in its burrow. Note the formal bedding of grasses. Penguins that live where grass or moss are not available, including all penguins that nest in the Antarctic, make nests of rocks.

the colony and parental care) of emperor penguins from large colonies is a migration of almost Serengeti proportions as thousands of juveniles depart from the colonies in massive groups over a week's time. From this great collective, I believe the groups break up into small groups or solo birds soon after departure from the colony. This would be to their advantage once on the high seas, where a large group would be attractive to predators but single juveniles or small numbers would be much harder to find and of less interest to predators.

There are massive departures of other penguin species, such as chinstrap, Adélie, king, macaroni, and little penguins at the Phillip Island observatory. To my knowledge, the dispersal and grouping behavior of these penguins away from shore is not well documented. I have seen king and macaroni penguins gathering on the shoreline, diving or dashing into the waves and gathering outside the breaker zone, and then disappearing. Again, I suspect that they depart in small groups to solo animals. "Crittercam" images from cameras mounted on the backs of penguins show other birds present when they are out to sea, but it is impossible to tell how many birds are in the group.

Then there are the more cryptic and small populations of penguins, such as the Galápagos, Humboldt, fiordland crested, yellow-eyed, and little penguin (in small colonies), which often leave and return at night; it is next to impossible to get a sense of their group behavior. Once penguins of any species go to sea, they are impossible to follow into the corners of their ocean habitat. Penguins remain an enigma in regard to their social behavior at sea.

Another unsettled question is what to call a group of penguins. We call a group of geese a gaggle, a group of quail a covey, a group of crows a murder, and a group of raptors a kettle, but what is the correct term for a group of penguins? On shore in the breeding area, it is a colony, but *colony* is appropriate for almost any breeding group of birds. What are they when gathered at the shore line? Maybe a mob; while on the ice, maybe a flock; and under water, perhaps a flight or squadron.

Do penguins fight?

Among perhaps all penguin species, with one exception, there is some degree of fighting. When mate selection is in process, there is mild, ritualized fighting and territorial behavior at the nest site. The bonded pair considers the nest and the reach of their outstretched necks and perhaps of their wings to be their turf. Any other bird that tries to enter this zone will be attacked vigorously; this prevents other pairs from disrupting their incubation, brooding, and feeding process. It also puts the pair on guard for thievery of their nesting stones, in the case of species that build a crude nest by piling stones in a ring to form the wall of the nest. Stone thievery is common among Adélie penguins (as shown on the BBC production of *Frozen Planet*) and among chinstrap penguins as well. At times, stones are scarce in a penguin colony, and thievery may occur in more species than just these two. It probably occurs in all species that place material in their nest. A female Adélie penguin of another mate will even lure a lone male protecting a nest site into sex, or at least away from the nest, and then depart with one of his nest stones. The nest site in most species is inflexible and a nest is rebuilt at the same spot in the colony, or near the same spot, from year to year. In the more flexible nesting sites of king penguins, the earlier nesters form the core of the colony, and as later breeders arrive, they form the cortex, or outer perimeter of the colony. This is a natural cycle for the king penguin that results from its extended breeding cycle, which lasts about 14 months. The early breeders in one year become late breeders the following year. A disadvantage for early breeders is that as the colony grows, they have a greater distance to go and more nests to pass by and through to reach their own nests. The path through the colony invariably is through the nesting territory of other pairs, and the transient bird incurs their wrath, pecks, and body blows from wings as it "runs" for the nest.

The building of a nest depression by king penguins represents little investment, and the parent with a single egg or chick may move to a different site, if its neighbors allow it, in the manner used by emperor penguins to move with an egg; they shuffle along supporting the egg or small chick on their feet while enduring beatings from their near neighbors. A move may be prompted by poor selection of a nest site, such as next to a glacial stream

Gentoo penguin stealing nest material from an incubator asleep on the job.

that rises during the summer melt or on the border of the beach where the nest is susceptible to unusual spring tides or high surf. Under these circumstances, king penguins have no choice but to force their way into the territory of other birds. Fighting over space becomes vicious and blood is spilled. In one case I saw an adult bird still brooding its egg although it was so badly injured that it was blind in both eyes and had no defense against the pecks and beatings from its near neighbors.

Strikingly different is the example of adult emperor penguins. The incubating males must form cooperative huddles in order to overcome the intense cold of winter. With the single egg or small chick supported on his feet, the penguin is free to move about with little or no interference from

King penguin nesters threatening intruders.

neighbors. Except for the occasional peck or wing slap, I have never (in 11 seasons of observations) seen emperor penguins fight while in the late brooding stages of their reproductive cycle. An interesting thought puzzle is, How small can an emperor penguin colony be and still be successful? If huddling is important for saving energy, then there must be a minimum size. Perhaps the colony that declined from 150 pairs in 1970 to nine males in 2005 and became extinct by 2009 had reached the minimum size for viability.

Do penguins bite?

Yes, most penguins are capable of biting each other with their bills and probably do so when circumstances warrant it. The heavy-billed crested and banded species, such as the macaroni and African penguins, are the ones most likely to bite. It is an offense least likely in the emperor penguin with its long and slender bill. King penguins do use the bill at times, but in their case stabbing is more effective than biting.

How smart are penguins?

A search through several books on penguins and other birds found none that deal with any intelligence measure—not surprising since there is much uncertainty as to what intelligence is or how to measure it. Most writings about animal intelligence focus on mammals, and the mammals most often discussed are apes, which have been extensively studied, especially the chimpanzee and the bonobo. With regard to marine mammals, the bottlenose dolphin, *Tursiops truncatus*, is thought to be very intelligent and has a brain-to-body mass greater than that of humans. Some people think birds are not intelligent and that much of their behavior is "hardwired," so to speak, or inherited. Yet, one of the largest brain-to-body-mass percentages is that of the American crow, which at 2.27% falls just short of the brain-to-body-mass percentage of humans at 2.5%, while that of the common raven is 1.33%. The raven is thought to be so exceptional that Bernd Heinrich devoted an entire book to it, titled *The Mind of the Raven*. His definition of intelligence is "doing the right thing under a novel situation." The use of tools is thought to be a reflection of intelligence and usually makes news when it is discovered in a mammal. However, there are examples of tool use in 104 species of birds. An exceptional bird that resolves novel problems with tools is the New Caledonia crow, which not only uses tools but also shapes them for the need. None of the birds in the long list are marine species.

Only two marine mammals are known to use tools, the sea otter and

Bottlenose dolphins, one of the few marine mammal animals observed to use tools. If penguins use tools, it will have to be in a different way from the way that dolphins use tools.

the bottlenose dolphin. The sea otter uses stones to break open shells and other hard prey. Dolphin tool use is only known to occur among some individuals, always female and only in Shark Bay, Western Australia. Those dolphins carry a sponge on the end of their snouts while searching for fish in sand bottoms; the sponge not only increases the area of disturbance during the search but also protects the end of the snout from abrasion. We know that killer whales that prey on other marine mammals and birds often work in coordinated packs similar to wolf packs. Their method of hunting and the creative nature of the tactics is an indication of intelligence. Other than these spectacular observations, most marine animals are out of sight during their aquatic activities, so any novel hunting behaviors or tool use is not likely to be observed. In short, we are still waiting to determine what kind of intelligence penguins have. They are clearly able to thrive in challenging environments, so it will not be surprising to learn one day that they have equal intelligence to their more northerly cousins.

Penguins are especially difficult to observe during foraging behavior. The oceanic species, which includes most of them, are seldom seen in the areas important to them for feeding. They also have very small brains relative to their body mass, with that of the emperor penguin measuring only 0.18%—not an impressive score. However, this oversimplified indication of intelligence illustrates the flaw in such measures. The body construction of the penguin is accentuated by a relatively large muscle mass, such that the two major chest muscles account for up to 75% of body mass. Few tetrapods (four-legged creatures with a backbone) have such an exaggerated emphasis on a specific muscle group. Penguins need the mass for strength and endurance and, as you will read in chapter 5, as an oxygen store. They are also not limited in size so much as flying species because they dwell much of the time in a gravity-free world.

In summary, we do not know how smart penguins are. Nevertheless, they are able to hunt and find prey in what is, to us, often a uniform environment with few landmarks as they travel hundreds to thousands of miles across vast areas of the southern seas.

Do penguins play?

Scientists have reported that only 25 species of birds engage in play and that the dominant groups among those are parrots and corvids. There are no reports of play for marine species. An observer of Adélie penguins in the 1970s thought they were "play chasing." The only other seabirds reported as playing are herring gulls and frigate birds. But both of these reports have been questioned as to whether the activities were really play. I have observed similar displays of exuberance in individual emperor penguins who were executing high-speed maneuvers: they created extensive bubble trails under water and engaged in somewhat similar vigorous exercise on snow-covered ice. There appeared to be no reason for the activities other than perhaps "loosening-up exercises." Emperor penguin chicks mimic these exercises when they are near the time of fledging (learning to fly); they will run with irregular tracks and simultaneously flap their wings vigorously.

Do penguins talk?

If to talk is "to convey information or communicate in any way (as with signs or sounds)," according to one of the definitions in the *Merriam Webster Collegiate Dictionary*, then penguins, like other birds, do talk. They communicate by means of several vocal and physical displays in breeding colonies. The combined vocal and visual displays are signals for contact, aggression, and sexual activity. The contact call and display in king penguins is usually stretching the neck upward, pointing the bill skyward, and trumpeting loudly, and then rapidly dropping the head and pointing the bill at the ground. There is a gender difference between the male and female vocalizations, which is identifiable to the human ear. Other information in the call identifies the individual and may reflect the robustness of the trumpeter. There are probably as many variations of this contact behavior as there are species of penguins, and the individuality of the call helps the female to find her former mate or select a new one. This behavior may not be so important in nest-building penguins as in the emperor and king penguins, which do not have formal nests or a specific location of their nests. In these two species, there is an added complexity of a two-voice system: both branches of the vocal organ (the syrinx) produce sound independently, providing a stronger signature of individuality. Later it is

Stereotyped posture of a king penguin during a contact call.

important, especially during the crèche stage (when the young are gathered apart from their parents), for a chick to recognize its parent and vice versa when the chick responds in kind to the trumpet call of the adult. The chick is very good at picking out the call of the parent amid the cacophony that is present in the colony when hundreds of other parents are calling at the same time. The chick is most sensitive to the frequency-modulated part of the call; it needs only to hear a small part of the call to recognize its parent.

Agonistic (aggressive) displays play an important role in communicating the mood and intention of a bird. There are three levels of behavior: (1) displacement activities, such as preening or wing-flapping; (2) submissive behavior, such as the exaggerated upright walk-away of the king penguin; and (3) aggressive behavior, in which the pattern is first a vocal and visual warning, with the back-of-the-head feathers of the Adélie penguin erect and the white eye ring prominent, and then strikes from the wings and pecks and bites with the bill. The more intense the vocal component of the display, the more likely it is that there will be a physical attack by the displaying bird. Both the visual and the vocal part of a display are important in the socializing between individual birds, and elements of the display are probably related to the quality or fitness of the putative partner.

Less dramatic but important, based on their frequency, are the contact calls. These are simple, brief, groanlike calls that adults are constantly emitting while in groups and when an external event occurs. Penguins standing at the ice edge and thinking about departing are very alert to any bird swimming by, and the event evokes numerous calls from birds both in and out of the water. A more dramatic call is the departure cry of juvenile emperor penguins as they walk to the ice edge. It is loud and distinctive and attracts other juveniles, which then join the parade to the sea.

How do penguins avoid predators?

Camouflage. As described in chapter 3, penguins all have a striking countershading pattern of black backs and white breasts with an extremely sharp boundary between the two. This well-defined boundary has little variation between individuals. The main advantage of this color pattern is in the water. If either an aerial or an aquatic predator is above a penguin, the black back blends with the aquatic background, especially if the bird is under water, and detection is difficult. If the aquatic predator is below the penguin, then the light background of the sky above the water surface disguises the silhouette of the white-breasted bird. Penguins with variant coloring, especially those that are completely black or completely white, must be at a disadvantage for detection by predators, and perhaps even for recognition by their own species.

Physical Avoidance. Penguins are exceptional aquatic athletes. Compared with other four-limbed marine animals, they can probably accelerate faster and make sharper and faster turns. They achieve maximum bursts of speed that are equal to or greater than the speeds of most seals and sea lions (pinnipeds) and some whales. My colleagues and I were able to observe the interaction between a leopard seal and some emperor penguins from both above and below the water surface. Over many hours of watching the birds as they arrived at the ice edge, we noted the agility and speed of emperor penguins under water. We measured a maximum swimming speed of 7 m/second, which was recently topped by penguins that another observer saw near the same site (see chapter 5). The 7 m/second record was the fastest reliably measured speed for any penguin at the time, and it occurred during a rapid acceleration from a depth of about 10 m as the bird ascended and rocketed out of the water onto the ice-covered surface.

Our conclusion is that no leopard seal is able to accelerate so rapidly or to that speed. Instead, the seal waits under water for penguins to depart

Emperor penguins clearing the ice foot freeboard. In order for the highest penguin to reach that height, it had accelerated to about 5 m/second.

from the edge of the ice, and it makes its capture during those few seconds when a bird is unable to see because of the stir of air bubbles created by the group departure. At that moment the penguin also enters a dark zone because of the brilliant light above the ice. Seals also reverse the tactic and wait on the ice for arriving birds. These methods of the seals are a matter of stealth; they do not usually chase a bird down in an open stretch of water. Surprise is probably used by leopard seals on other species of Antarctic or sub-Antarctic penguins as well. And it is most likely the strategy that fur seals use for capturing penguins that live in temperate waters.

Perhaps the greatest potential threat from the sea is from the "wolf pack" strategy of killer whales. They use various methods in their pursuit of individual birds or groups of penguins. Until several years ago, killer whales had only been observed preying upon king penguins in the Crozet Archipelago. Then in 2009, in the Antarctic Peninsula, killer whales were observed catching and eating gentoo and chinstrap penguins. The sophisticated method of stripping only the breast muscles from the carcass suggests that the predator is experienced in this type of feeding.

There are no aerial predators of penguins that are at sea. It seems possible that sea eagles might take some of the smaller species, and there is a remarkable confirmation of this for the little penguin. A satellite transmitter that was attached to a little penguin began sending signals from high on a cliff. On investigation, an Australian national park employee found the transmitter in the nest of a sea eagle. Whether the little penguin was captured on shore or at sea is unknown. On land both the coloration and the physical skills of penguins are of marginal help to them. An array of gulls, eagles, and caracaras are potential predators of temperate species, especially of the young. The greatest protection for the adults is avoidance

Adélie penguins departing to sea as a group. This safety-in-numbers strategy of predator avoidance is a common behavior for both Adélie and emperor penguins as they depart their colonies.

by choosing a good location for arrival to and departure from the colony. Most penguin colonies are on islands where there are no land predators, except for introduced species such as rats and cats. The greatest threat to island and Antarctic penguins are aerial predators that feed upon the chicks in the colonies. All Antarctic and sub-Antarctic penguin colonies are exposed to predation by brown and south polar skuas and giant petrels. As for the example of the little penguin and the remarkable coincidence of a satellite-tagged bird being taken, sea eagles may be an important predator and may be one of the reasons this species almost always comes ashore at night.

Vigilance. Penguins, at least emperor and Adélie penguins, exercise a high degree of vigilance when they depart from ice edges near their colonies. The careful observation by the birds at the ice edge usually results in a crowd waiting to go. On some unknown cue, many birds, if not the entire group, depart. These departures can be spectacular events when a large number leave: the phenomenon reminds me of the rush to board the train in Tokyo during commuter hours. The largest departure I ever witnessed was a mob of about 4,000 emperor penguins all leaving from a narrow section of the ice edge. It took over four minutes for all the penguins to leave.

There are many iconic photos of Adélie penguins departing from a high ice ledge. This strategy in various forms is one of "safety in numbers." The birds swamp the individual leopard seal that may be lurking below, so that in a single departure only one capture occurs.

Sociality. During the waiting period at the ice edge, there is muted interaction among the birds, such as occasional pecking and wing chops at the back of a near neighbor or stretching of the neck to look over the front part of the crowd and out to sea. When birds come swimming past the edge, they attract much attention from the birds ashore, and there is much calling between the birds at sea and those on land, as described earlier. It is almost as if they are asking whether there is a leopard seal skulking nearby. Once the birds are in the water, they seem to gather into a group as they do preliminary exercise maneuvers before setting off on their distant goals. If a leopard seal or a killer whale shows up, the rule seems to be every bird for itself; they scatter in all directions including back onto the ice. All of this behavior may occur among chinstraps and gentoos when they are doing an on-ice departure. Group behavior of penguins in more northern latitudes, where there is no ice from which to watch, is less clear. In the far south, the ice edge is a wonderful platform for bird watching that very few people have the opportunity to experience. I suspect that on most tourist expeditions, the tourists rush past one of the greatest shows on earth to get to the routine life playing out at the colony.

Chapter 5

Penguin Ecology

Where do penguins sleep?

The sleeping habits of penguins depend on the season. During the reproductive period at the stage of courtship, the pair may share the nest on the surface of the land, in burrows, or in a cave. After the eggs are laid, while one sleeps on the egg or eggs, the other is usually off at sea foraging. During extended periods at sea, what the birds do at night is uncertain. They may sleep and drift on the high seas, or if they are polar penguins, they will sleep on ice floes in the pack ice. Time-depth records from king penguins indicate they sleep little while on foraging trips between brood and feeding intervals, since they dive constantly during the day and night. The sea trips are especially demanding as they search for prey for about one to three weeks.

During the winter, king penguins travel to the marginal ice zone, where after a considerable trip across open water, they may have an opportunity to rest; these trips last three months and reach distances of at least 1,600 km from their colony. Whether they rest is unknown, because foraging activity has not been determined. As for polar penguins, the female emperor penguin, on long journeys while the male incubates the egg, may sleep on ice floes in the pack ice where she is foraging. Time-depth recorders attached to the females show that they are out of the water most of the time and resting on ice floes. When emperor penguins, either male or female, return to the colony after foraging trips, they are commonly seen lying down and sleeping a short distance from the aggregation of chicks, which by this time are in the crèche, or "day care," stage. At a similar time in the breeding cycle of king penguins, adults often sleep on the beach before

African penguins sharing the entrance to their underground nest.

departing the colony for their long foraging trip in the open sea. Overall, while in the colony attending to their egg or chick responsibilities, king penguins sleep a maximum of about 9% of the time toward the end of the breeding fast and only 0.3% of the time soon after the chick hatches. In comparison, most humans sleep or rest about 30% of the time.

For the smaller penguins, while nurturing chicks, their time at sea is usually for the day, sometimes overnight, and occasionally for several days. The period of time away is usually so short that it is unlikely that they sleep much while at sea; any sleep they get probably occurs while ashore. During the nonbreeding phase of the annual cycle, however, many species of penguins migrate and remain at sea for long periods of time. While on the high seas during this nonbreeding period, they probably sleep on the water during the night, perhaps experiencing unihemispherical sleep (when half of the brain is awake).

Do penguins migrate?

Almost all penguins migrate. Migration here is defined as traveling to a distant place outside the breeding season; it includes travel to a molting area and both pre- and postmolt winter travels to foraging areas. Not included are foraging trips taken while raising chicks and the virtually unknown travels of juvenile and immature birds. For example, the startling discovery of a juvenile emperor penguin arriving on Peka Peka Beach, New Zealand, in June 2011 would be considered extreme, even though most emperor penguin juveniles do "wander" considerable distances that are out of the ordinary for adults. In fact, juveniles leaving their natal colony continue beyond Antarctic waters during the summer and do not return to the Southern Ocean until the next winter. Their ventures for the next two or three years, like those of other penguins at this age, are unknown. The

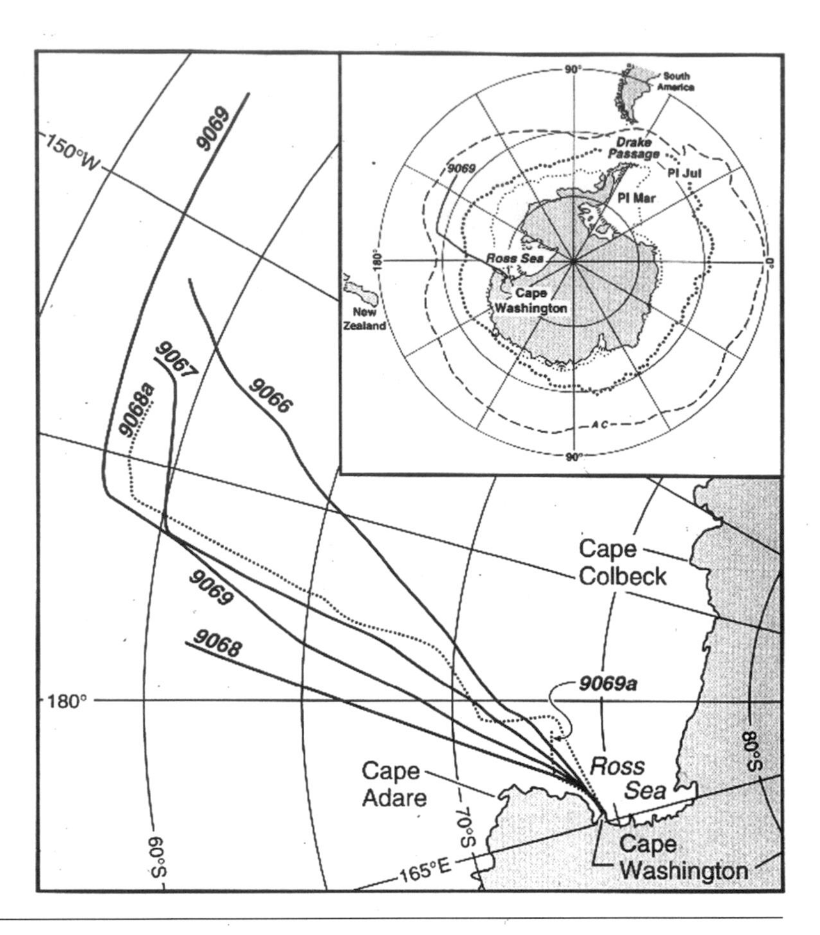

Emperor penguin juvenile tracks after departure from the Cape Washington colony over three sequential years, 1994 to 1996. The longest straight-line distance from the colony was of bird 9069, which at last transmission was 2,845 km from the colony. Figure modified from Kooyman et al. 1996.

Galápagos, Humboldt, little, gentoo, and yellow-eyed penguins usually do not migrate. An outlier is one Humboldt penguin. Out of five tracked, this one traveled north about 640 km from the colony, well beyond the 35 km protection zone that exists around some breeding colonies. However, this is a modest distance compared with species that are routine migrators.

Among the true migrators from the cold-temperate regions are adult birds that have been studied during the postbreeding season. One of these, the short-crested rockhopper (*Eudyptes c. chrysocome*) breeding at Staten Island, Argentina, travels less than 1,000 km from the colony, but one exception, out of 24 birds studied, traveled more than 2,000 km from the colony, as measured in a direct line. The important wintering areas were off the north coast of Tierra del Fuego and Burdwood Bank south of the Falkland Islands. Remarkably, some spent time in Drake Passage, one of the most violent areas of the world's oceans. Magellanic penguins of the Falkland Islands prefer milder seas and may migrate northwest to northeast, reaching distances of 1,800 km from the colony. Those that migrate to the northwest travel along the northern Argentinean coast. Magellanic penguins breeding on the mainland of northern Argentina also migrate to

Table 5.1. Pre- and postmolt days away from the colony and maximum depth measured during foraging of oceanic and nearshore foragers while nurturing chicks.

Species	Premolt duration (days)	Postmolt duration (days)	Maximum depth (m)
Oceanic foragers and travelers, often out of sight of land			
Emperor penguin	30	60	564
King penguin	30	ca. 30	348
Adélie penguin	ca. 30	210	180
Chinstrap penguin	n.d.	210	179
Magellanic penguin	n.d.	180	97
Macaroni penguin	14	180	163
Royal penguin	30	180	226
Erect crested penguin	30	130	n.d.
Snares penguin	70	120	n.d.
Northern rockhopper penguin	50–60	180	168
Nearshore foragers and travelers, often within sight of land			
Gentoo penguin	30–60	n.d.	212
African penguin	n.d.	n.d.	130
Humboldt penguin	n.d.	n.d.	54
Galápagos penguin	n.d.	n.d.	32
Yellow-eyed penguin[a]	n.d.	n.d.	56
Little penguin	n.d.	n.d.	67

Notes: n.d. means no data. For the Fiordland penguin, there are no data at all on its foraging habits.

[a] The only known consistent bottom (benthic) forager.

the north and sometimes reach the southern coast of Brazil, about 1,800 km to the northeast.

The well-studied travels of the Adélie penguin show birds traveling from Signy Island, South Orkney Islands, to a molt site in pack ice that is up to 1,000 km south. On the other side of the Antarctic, the route of satellite-tagged adult birds has been tracked after molt from Bechervaise Island. For five months the birds traveled west before moving north with the expanding pack ice. The maximum distance noted from the colony was 1,600 km. And an Adélie penguin, during its winter migration from a colony deep in the Ross Sea, migrated to the Southern Ocean north of the Ross Sea and the Balleny Islands. This was a nearly direct distance of about 1,500 km. Is there a pattern among these various studies of penguins? Perhaps so, and patterns probably pertain to particular species and colonies, but the data are too limited to support anything but the broad conclusion that the offshore species leave the area of the colony for an extended period and travel considerable distances.

One of the most extraordinary journeys, and one that is probably consistent among king penguins but needs to be verified among other colonies, is the travel by king penguins from Crozet Archipelago to the marginal ice zone of the Antarctic during the winter. Sometimes they were tracked to more than 1,900 km from the colony, and they must return eventually to care for the chick that is over-wintering. It is the only species in which the chick does over-winter before fledging (going to sea) in the following spring. In a sense this is a foraging trip of several weeks' duration. If categorized in that way, this trip by king penguins is the longest-duration foraging trip, while nurturing a chick, of any penguin species.

The king penguin's closest relative, the emperor penguin, takes an extended premolt journey from the colony to the molt site. Emperor penguins leaving the western Ross Sea colonies take a month to travel about 1,200 km, during which time they gain considerable weight in preparation for the monthlong molt. After the molt, they take about two months to return to the colony for the next year's reproduction cycle. The emperor penguin's and the king penguin's postmolt absences before returning to the colony for reproduction are shorter than those of all other oceanic penguins. These two species are constantly on a fast track, so to speak, to raise their chicks.

Data for nearshore forager migration patterns, if nearshore penguins have any such patterns, is lacking, with the exception of the one Humboldt penguin mentioned earlier that traveled 640 km from its colony.

Are penguins good divers?

A hint of how good penguins are at diving is that they are the only group of birds whose wings are totally committed to underwater flight and that their wings are often called flippers rather than wings. Two of the most fundamental criteria for good divers are how deep they dive and how long they stay submerged. These are relative terms in regard to other divers, and there are a variety of avian divers to which penguins can be compared. These can be divided into nearshore, or neritic, foragers, meaning they remain over the continental shelf at depths from the low-tide mark to 200 m, and pelagic or oceanic foragers, which dive in deeper waters. There are about 122 species of birds that dive, spread through eight different families. The most diverse among these divers are the cormorants, with 27 species within the family. The least diverse are the loons and diving petrels, with four each in their families. The smallest are in the family Alcidae, the auklets, murres, and puffins. Historically, with the exception of some penguins, the largest avian diver was the flightless great auk, the last breeding pair of which died in 1844, when they fell victim to collectors.

A fat king penguin in the obese condition suitable for the long fast during the molt.

Dive-duration data exist for few of these avian divers because, since most are small flyers, they could never be burdened with time-depth recorders, which at a minimum weigh about 25 g and are too large for most species. Current records for diving depths are obtained with capillary depth tubes, which weigh about 5 g and yield only the maximum depth for the entire foraging trip. Many diving birds are not studied because they are inaccessible, occurring on remote islands or coastlines and often on sheer cliffs.

All these variables considered, it is remarkable that one of the deepest dives on record is for a bird other than penguins, the oceanic Brünnich's, or thick-billed, murre. Obtaining the data for this dive was heroic, because the investigator had to rappel down a sheer cliff more than 100 m high to capture the bird and attach the capillary depth recorder. Then a day later he had to do it again to recover the recorder, which showed a maximum depth of 210 m. All murres are flyers, but they use a partially folded wing for propulsion under water. They could easily be mistaken for penguins, when resting on a cliff roost or swimming under water, except that they are in the wrong hemisphere. Since they fly both above and under water, they must be small. None weigh more than 1 kg, about equivalent to the smallest penguin's weight. Yet the little penguin's deepest recorded dive is only 67 m. This is one of the differences between the pelagic (oceanic) and neritic (nearshore) divers. The latter may not have the incentive to dive as deeply as the pelagic foraging murre. In contrast, the next-deepest record for a nonpenguin diver is the nearshore Crozet shag (145 m), which is often a bottom feeder. The records for the shag and the murre are also remarkable compared with the records for penguins in general. There are few penguin maximum depths that exceed those of the shag, and the only dives deeper than the murre depth are by four of the largest penguins, the royal, the gentoo, the king, and the emperor penguins.

With the exception of flightless outliers in the family of alcids (the great auk, now extinct), the Galápagos Islands' flightless cormorant, and the Tierra del Fuegan flightless steamer duck, penguins are the only group in which all of the birds are flightless. The age of this group and its diversity are discussed in chapter 1. Other flightless birds use foot power; in the case of the great auk, it is assumed that, like all other auks, it used wing propulsion. However, although the auk has a stubby wing that works for flying and diving, it is not different to the degree of the wing or flipper of the penguin; it is not as good for diving and also is a compromise for good flying. The penguin wing is rigid, with very limited flexibility. One of the most recognizable features of the penguin skeleton, besides the conversion of the flying wing into a flipper, is the pachyostotic skeleton, referring to increased bone density. These modifications were all in place by the late Cretaceous or the Paleocene period.

Size is important in divers, and so the closeness of the gentoo maximum dive depth to that of the Brünnich's murre is surprising, both because the murre is so much smaller than the 8 kg gentoo and because the murre is a flying species. Because the gentoo penguin is a neritic forager, it lacks the incentive of the pelagic-feeding murre. In an assessment of the diving depths of the two largest, and pelagic, penguins, the king and the emperor penguin, both size and their offshore feeding expose remarkable capacities. At a maximum recorded depth of 564 m, the emperor penguin is far and away the greatest of all avian divers. Any question of its capacity as a stellar diver is answered by the almost impossible-to-believe dive duration, for one exceptional effort, of 27.6 minutes. And it performs feats like this while diving under ice. No sane emperor penguin would intentionally make such a long dive. The long duration of the emperor penguin's dive probably developed as a result of diving under pack ice, at a modest depth of 110 m, when the lead or hole that it dived from closed before it returned, forcing it to search for another breathing hole. This penguin's capacity challenges even some of the best of marine mammal divers. No dolphin, fur seal, or sea lion has been recorded making such a long dive, and few dolphins, baleen whales, fur seals, or sea lions have been recorded as diving so deep.

Table 5.2. Maximum diving depths of divers other than penguins.

Species	Number of species in group	Maximum depth (m)
Nearshore foragers (flight and foot propulsion)		
Loon		4
Common loon		55
Grebe	20	n.d.
Sea duck	18	n.d.
Cormorant	27	
Crozet shag		145
Oceanic foragers (flight and wing propulsion)		
Gannet and booby	9	
Blue-footed booby		22
Shearwater	18	
Sooty		70
Diving petrel		
Peruvian	1	83
Auk (auklet, murre, puffin)	22	
Rhinoceros auklet		60
Atlantic puffin		68
Brünnich's murre		210

Note: n.d. means no data.

The case rests, that emperor penguins are one of the greatest diving species on the planet and that diving birds as a group are exceptional divers, especially if their size is taken into consideration. Size does matter. The larger the diver, the lower the metabolic rate, proportional to the oxygen store, that the lungs, blood, and muscle provide. These factors are important for breath-holding. The critical factors for the limitation of depth are not known.

In thinking about the importance of size, it stretches one's imagination to think what the great auk, at 5 kg, about five times the weight of the Brünnich's murre, was able to do. It requires even more imagination to guess what the giant penguins of the Eocene period were hunting and how deep they dived to find their prey. A giant penguin probably weighed about five times the mass of the present-day emperor penguin.

How fast do penguins swim, and how fast can they swim?

Both the questions of "do" and "can" are almost impossible to determine, but a few satisfactory measurements have been made. Indeed, the only maximum speeds on record were obtained during an elegant analysis of emperor penguins accelerating from depth to obtain high exit speeds that would launch them out of the water and over an ice ledge to land on fast ice (the unmoving ice along the shore). The authors of this report proposed that bubble clouds (I call them contrails, because of the visual similarity to those created by high-flying aircraft) are a mechanism for drag reduction by "air lubrication," enabling the birds to reach a maximum speed of about 8.2 m/second. Are such speeds comparable in other penguins? We do not know, because no maximum-speed tests have been conducted on other penguins. I can vouch from personal observations that contrails do occur when Adélie penguins are rising rapidly to the surface and getting ready to leave the water and land on fast ice. They probably occur for most if not all other penguins as well when there is a need to leave the water at high speed. Even short-term accelerations to high velocity with assistance from bubble release would be helpful in escaping from an aquatic predator in pursuit.

For the "do" part of the question, most penguins cruise at about the same speed of 2.0 m/second, or 7.2 km/hour. *Cruise* is defined as the speed measured while diving and searching for prey, and there have been only a few such measurements obtained. As for times when there is a need to travel faster than the diving cruise speed, for example, when they are departing or returning to the colony, they often porpoise (leap or plunge like a porpoise). There are measurements of porpoising speed for Adélie penguins that are at 2.8 m/second. This is just a bit lower than the calculated porpoising speed of the 5 kg macaroni penguin, whose body mass is equivalent to that of the Adélie penguin. Among penguins, the emperor penguin is the only one that does not porpoise, but more than 50% of the front half of that penguin's body often rises well above the surface while the bird is taking a breath.

Which regions have the most species of penguins?

There are nine species of penguins, mainly crested penguins, that live in the complex of islands within the latitude of 45° to 60°S and from the longitude of 170°W to 145°E. This includes the islands of Chatham to New Zealand to Tasmania in the north, to Macquarie in the south, with Snares, Auckland, Campbell, Bounty, and Antipodes Islands in between. These are cold-temperate waters of the Pacific, and in this region the diversity of the crested penguins makes up the majority of the species. None of the banded penguins live here, and the *Spheniscus* are as widely separated from each other as the continents where they occur. The one exception is the Galápagos penguin. Despite its occurrence on the equator, the habitat is cold-temperate and similar to that of the other banded penguins.

Table 5.3. Swimming speed of penguins.

Species	Cruising (m/s)	Porpoising (m/s)	Maximum (m/s)
Adélie penguin	2.0	2.8	
African penguin	1.8	n.d.	
Emperor penguin	2.1	4.4[a]	8.2[b]
Gentoo penguin	n.d.	3.4	
King penguin	n.d.	3.9	
Little penguin	0.7	2.6	
Macaroni penguin	n.d.	3.2	

Notes: n.d. means no data. Porpoising speeds are all calculated figures, except for the Adélie penguin's, which was measured directly.

[a]Calculated speed necessary to porpoise. I have never seen emperor penguins porpoise, in hundreds of hours of observation.

[b]Calculated from swim velocities obtained from underwater video as birds accelerated to leap up onto ice.

Reconstruction of the giant penguin *Anthropornis nordenskjoeldi* with living emperor penguin, chinstrap penguin, and little blue penguin for scale. Art by Kristin S. Lamm. Modified from Ksepka et al. 2006.

How do penguins survive the extreme cold?

In this case, extreme cold occurs mainly in the Antarctic, where temperatures at some emperor penguin colonies may drop, for short periods of time, to –58° to –76°F (–50° to –60°C) and continuously hold at about –40°F (–40°C) for weeks at the most southern colonies. Added to the extreme cold is the wind, which, when converted to a wind chill factor, indicates the full impact of cold on the potential for freezing bare skin. There are two ways of adapting to the cold. One is anatomical and physiological adaptation, and the other is behavioral adaptation. The most extreme example of each is exhibited by the emperor penguin, the most cold-adapted of all penguins. It breeds during the Antarctic winter and is therefore forced to adapt to some of the most severe marine weather conditions of the Antarctic. During the breeding season of late autumn, the males and the females meet on the fast-ice surface near the continent. At this time temperatures may drop to as low as –40°F (–40°C) and winds in the colony may reach 80 km/hour. Under these conditions, the wind chill factor brings the effective temperature to –88.6°F (–67°C). This temperature is very dangerous for most animals, especially humans, which are a tropical species by origin. Considering that the thermal-neutral zone (that temperature when the metabolic rate must increase above resting to maintain body temperature) of emperor penguins is 14°F (–10°C), this means that without some behavioral adaptation, their energy consumption and the winter temperatures of their breeding habitat would be intolerable. After laying the egg, the female departs to forage, and the winter temperatures sink even lower; subsequent storms may produce even stronger winds. Meanwhile, the male hunkers down to incubate the egg and to endure what the winter offers up.

The physique of the emperor penguin establishes it as the species with the highest tolerance to cold. Because it is the largest penguin, its surface-area-to-body-mass ratio is the lowest and most favorable for reducing heat loss in air or water. Additionally, this penguin has a longer and denser feather coat than any other penguin species, with the most complete coverage. The only bare skin is the large feet, but except for an area on the heel about the size of a U.S. quarter, none of the foot touches the snow or ice while the penguin rests and stands in the tripod configuration of the two feet and the stiff tail feathers. The foot is also tucked up under the feathers, and in this posture it is favorably insulated. No other penguin can rest while standing and still cover the feet, because their abdominal feathers are not long enough. Also, as the bird rocks back into the tripod stance, the small surface area on the heel of the foot minimizes the amount of body contact with the ice and heat loss.

A further anatomical adaptation is the countercurrent circulation system in the foot. Countercurrent circulation is a common adaptation for conserving heat by reducing heat loss in extremities in penguins and many other birds and mammals. In the case of the penguin foot, it can be many degrees cooler than the body core. Arterial blood at core temperature flows to the foot, and the adjacent cold venous blood returning to the body core warms as heat is transferred from the arterial to the venous blood. This vascular adaptation reduces by a great deal the overall heat lost by the blood to the foot, compared with what it would be if there were no

Emperor penguins in autumn.

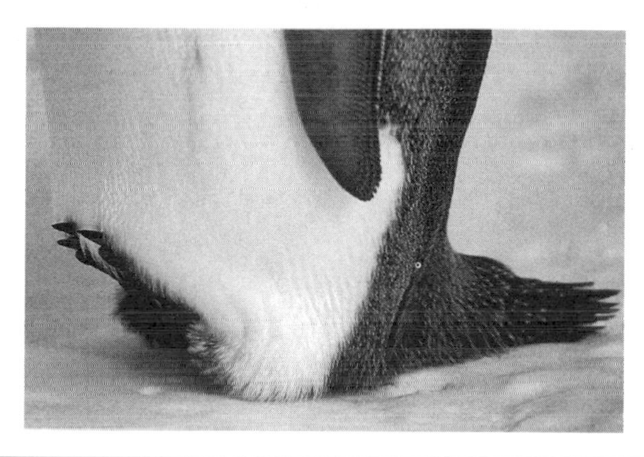
Emperor penguin in the tripod stance with the feet covered.

countercurrent flow. The continued flow to the feet is necessary to provide enough circulation and heat transfer to keep the feet from freezing, as well as to supply oxygen and nutrients to the foot tissues. A similar anatomical arrangement is necessary for circulation to the wing. This heat-conserving arrangement of a complex network of arterioles and venules is found not only in the wings and legs of penguins, but in the flukes and flippers of whales and seals and in many other warm-blooded birds and mammals. These networks and countercurrents are such effective modes of biological engineering that they are common for other functions as well. There are countercurrent systems in some birds and reptiles that have extraorbital salt glands, in the kidneys of mammals, and in the gills and swim bladders of fish, to mention a few. They are part of the salt-gland structure in birds and reptiles to help maintain electrolyte balance in their blood. Ever wonder how the swim bladders of some deep-dwelling fish maintain volume under high water pressure? It is because of a countercurrent blood-flow system and gas transfer from the blood to the swim bladder. To learn more about countercurrent systems, read Per F. Scholander, *The Wonderful Net*, or Knut Schmidt-Nielsen, *Animal Physiology*.

There are also behavioral means of preserving the body temperature, and again, no other animal needs such strategies more nor is more effective in their use than the emperor penguin, especially the male while he incubates the egg. The males have no other choice than to remain exposed to the harsh conditions of winter, and along with physiological and anatomical adaptations, they enhance their thermoneutral zone by clustering into dense groups called huddles during the most severe weather. Huddling enables them to reduce their metabolic rate below the basal metabolic rate. During extreme cold, which usually means a blizzard, group cooperation has great importance because of the huddle. This activity is so effective that birds deep in the interior of a large huddle of tens to hundreds of birds may become too warm. Surface temperatures of birds within the huddle sometimes reach near 95°F (35°C), which is too hot for emperor penguins, and they move to the periphery or even outside of the huddle to cool off. Huddling is a very dynamic process; the birds break up frequently on average every 1.6 hours. Males huddle only about 38% of the time during incubation. Adélie penguins may also huddle during the winter when large aggregations occur in the pack ice. It is a guess, but king penguins that winter in the marginal ice zone may huddle as well. Individual penguins traveling alone, or while they are at the colony but not associated with other penguins, lie down on the snow and tuck their wings and feet under them the best they can. If during this rest there is blowing snow, they are soon buried and benefit from the protection of the blanket of snow. An indication of how effectively their feathered coat insulates them is that the snow covering does not melt.

How do penguins survive the winter?

The previous answer describes how male emperor penguins spend the winter. As for nonbreeding males, females, and all other penguins, winter is a time of unstable temperatures and mild to violent seas in which they spend most if not all their time. There is also a gradient in winter conditions from the severe weather of the south polar seas to the mild, idyllic weather and seas of the Galápagos. Polar penguins are usually in pack ice, which calms the seas. However, there are long periods of darkness, and for all penguins studied, observations show that they rest on the pack ice at night and enter the water only during the daylight hours. The species that live farther north and go on long winter migrations feed during the daytime and travel or rest on the land or ice surface at night. The nonmigratory species forage daily and usually rest ashore at night. For all species winter is a challenging season because the summer and fall abundance of prey often decreases. The penguins must continue to find food, even though a more modest amount than when they are nurturing chicks. Nevertheless, sometimes they are unsuccessful in sustaining themselves, based on reports such as those of Magellanic penguins found in poor condition along the coast of Brazil.

Do penguins get too hot?

While resting in water, all penguins rely on the waterproof insulation of their feathers to prevent heat loss across the large temperature difference between the water and their bodies. They will not get too hot. When the penguin is swimming vigorously, the exceptional insulation provided by the feathers now is a barrier to the transfer of the additional heat produced by the activity. The penguin is like an enclosed room with a heater, where it gets too warm when the fire is turned up. In the room, the windows can be opened and a fan turned on to dissipate the excess heat. The "windows" of the penguin are the thinly covered wings, bare feet, and facial skin. The "fan" is increased circulation carrying hot blood from the interior of the body to the face, feet, and wings. With the increased convection of water flow over the swimming bird, its body temperature is maintained through these heat-conducting windows. That is why the pale feet of rockhopper or Adélie penguins are pink when they first come ashore.

On land, thermoregulation is another matter. The thermoneutral zone of the emperor penguin is 14° to 68°F (–10° to 20°C), close to that of the Adélie penguin, which ranges from 14° to 86°F (–10° to 30°C). In such a cold environment as the Antarctic, that upper temperature is only a small problem, except for a short time in the late summer. In that situation, and especially for the more northern-distributed individuals of the Antarctic Peninsula, the Adélie penguin pants, but at a low increase in metabolic heat production. It is a matter of degree for the Humboldt penguin, which breeds along the shores of Chile and Peru. There the hot period can be months long and the ambient temperatures much higher, under a desert sun with no freshwater. Increased blood flow and heat loss through the bare facial skin, in addition to the feet and wings, is probably an effective means of maintaining body temperature during hot periods while ashore, but to my knowledge the thermal response of Humboldt penguins has been little studied.

Do penguins have enemies?

The *Merriam-Webster Collegiate Dictionary* defines an enemy as "one that is antagonistic to another; *especially:* one seeking to injure, overthrow, or confound an opponent." According to that definition, penguins, much like humans, may be their own worst enemies. With the exception of the emperor penguin, which relies on group cooperation for successful breeding and has no nest to defend, penguins can be very antagonistic toward one another. During nest defense, the attacks can clearly seek to injure the opponent. Depending on the species, its geographical location, and the type of nest, such as open ground, cave, or burrow, the intensity of fighting will vary. Penguins also have an array of predators. Emperor penguins, in the most southern part of their range, have the fewest predators. The only predator in the far south is the leopard seal; the killer whale has also been recently noted to kill penguins in the Antarctic Peninsula. In the northern part of emperor penguins' Antarctic and sub-Antarctic range, giant petrels, south polar and brown skuas, and an array of fur seal species can be added to the list. The chicks and the recently fledged juveniles are the most vulnerable to most of these predators. For chinstraps and gentoos in the Antarctic Peninsula, the predators are the same, in addition to killer whales, which have been observed recently taking these two species. The variety of predators is increased for the more northern species. Humans were once an important predator wherever they overlapped in range. Since most penguin species in the northern regions are now considered endangered or threatened, they are protected from hunters and eggers. However, for those species that live on continents, such as the Magellanic, African, and little penguin, almost any predator that hunts coastally is a potential enemy; those predators range from eagles to lions and include the household pet cats and dogs, as well as rats. Remote island species have been the least vulnerable, because of the few predators that occur naturally in these island groups. That was true until most of these island groups were subjected to species introduced by explorers and settlers. Since then the invaders have taken a heavy toll. In some of the offshore islands of New Zealand, South Georgia Island, and Galápagos Island, there are active programs to eliminate the introduced species. For more information about such programs, see the Island Conservation website at www.islandconservation.org.

Do penguins get sick?

All animals are at risk of infection by a variety of carriers, which range from direct exposure to infected members of their own species to vectors such as those residing on the body, such as lice, ticks, and mites, and airborne biting flies and mosquitoes. In my experience with mostly subpolar and polar species, I have seen few ectoparasites, or exterior parasites. The most obvious was a heavy infestation of ticks on the backs of the heads of some king penguins in the Crozet Archipelago. Any such vectors put the birds at risk for infection. One of the most common means of determining whether a penguin has been exposed to a disease is by a blood scan for antibodies, and a variety of antibodies for viral and bacterial diseases have been isolated from penguins. Of special concern are the Antarctic penguins because of their isolation and the presence of few if any endemic diseases; it is thought that they are especially susceptible to diseases that are brought into the region. So far there are few antibodies associated with diseases in penguins, but in some of those cases the antibody presence is very high and is a cause of concern. For example, one report from the Australian Antarctic Division says that antibodies for infectious bursal disease virus were found in 93% and 100% of the emperor penguin chicks at the Auster and Amanda Bay colonies, respectively. This seems like bad news, but up to the present, mass die-offs of penguins are uncommon; most deaths are a result of starvation rather than disease. As a precaution related to tourist travel in the Antarctic, which is steadily increasing every year, there are regulations for the use of prepared poultry at the stations and in the field camps. For example, any poultry taken into the field by U.S. projects must be skinless and boneless, and all wrappings are returned to the United States for disposal.

How can you tell if a penguin is sick?

Most penguins that are not well are found at the edge of a colony, or near the shoreline, and they are not moving to or from the colony in the pattern followed by other birds. Ill birds may also come ashore in unusual places, not near a colony, and sometimes in unexpected places outside of their usual range. A third clue is the appearance of these birds. If they are sick, they are likely to be weak and emaciated. Thin penguins are most commonly a result of starvation rather than disease.

The degree of underweight in a penguin can be judged by the prominence of the breastbone ridge. Other signs include conspicuous shoulder bones. The root cause of the bird's starving is seldom resolved, because it would require blood and tissue sampling and later analysis in a laboratory. In my own experience at polar penguin colonies, dead adults found near colonies that do not show some external trauma, such as a bite wound, are exceedingly rare, and birds that are obviously sick are uncommon.

Starved and emaciated king penguin chick.

Are penguins good for the environment?

As a predator at or near the top of the food chain, or food pyramid, penguins play an important role in the balance and richness of the ecosystem of which they are a part, depending of course on how abundant they are. Their contribution to the environment as a predator of small and abundant fish and as prey of other predators adds to the diversity and health of the environment and ecosystem in which they occur.

Chapter 6

Reproduction and Development

How do penguins reproduce?

Like all other birds, penguins must lay eggs to reproduce, and there is a complex series of conditions and rituals that lead up to egg laying. Reproductive behavior is similar among all species of penguins. There is a ritual of some kind for selecting and identifying a mate, followed by or concurrent with the selection of a nest site, except in the emperor penguin, which has no nest. Then, after a varying length of courtship behavior, a mate is identified and a bond is formed by visual and sound displays of posturing and trumpeting. A few days after copulation, egg laying occurs. Based on the time between egg layings, egg formation takes about three to four days. Most species have patterns of synchronous breeding (breedings separated by only a few days to a few weeks), long fasts during the incubation period, and intense foraging efforts after eggs hatch and during the nurturing of chicks. A little odd is that in some species of crested penguins, the two parents share the first period of incubation for about 10 days. This seems unusual because after the courtship and laying fast, one of the pair should go to sea to recover from that fast and prepare for the long fast of incubation. All penguins are monogamous for the breeding season, with a few exceptions of infidelity, but from year to year fidelity varies from the lowest in emperor penguins, in which about 15% have the same partner the following year, to 82% in the little penguin.

Where do penguins lay their eggs?

Penguins lay their eggs in a variety of places, ranging from burrows, to rocky or sandy substrates, to open spaces on fast ice (the ice attached to the shoreline that does not move). The investment in time and effort is probably greatest for nest builders that burrow. They have to dig a hole if none is available from the previous year of nesting. The nest inside a burrow is lined simply with a little grass, if anything. The little penguin is the most committed to burrow nesting or other means of undercover shelter for defense against predators. The Humboldt and Magellanic penguins at times have a mix of below- and aboveground nests in the same colony. Even above ground, the nests are not much more than a depression with a few strands of grass. The most elaborate nests are those of the brush-tailed penguins, all of which add a large number of pebbles to their nests. These rocks function to raise the nests, and during the summer thaw, this practice keeps the nest dryer than a simple depression would. King penguins vary from a simple depression in sandy surfaces to not even a depression on rocky surfaces. If conditions become unsuitable while parents incubate the single egg or small chick, they will move. The emperor penguin nest is the top of the parents' feet and the brood pouch. They are always on the move, balancing the single egg or brooded chick; the small chick is held within, or against, the incubation patch.

Magellanic penguin above ground with egg and first hatched chick. Though above ground, it is not entirely exposed. Note the bush in the background extending out over the nest.

Chinstrap penguin nest in a lake of thawed snow. In unusual years of much snow and flooding, the use of rocks to elevate the nest and eggs is essential to breeding success.

Does a penguin nest at the same time and in the same place every year?

Most species of penguin are synchronous in their breeding, but there are three types of synchronous breeding. Spring breeders include most species in the *Pygoscelis* and *Eudyptes* genera. An exception is the gentoo penguin; some of them may breed in the winter. Synchronous winter breeders are the emperor and the fiordland penguins. Then there are the synchronous but random breeders, which may breed any time of the year. These are the African, little, Galápagos, and Humboldt penguins. The most variable are the Galápagos penguin and the Humboldt penguin. Those that are so irregular live in the most variable environments and seize the moment when conditions are propitious for breeding success.

The spring breeders are the most synchronous, because they have the most reliable marine resources and are influenced the most by photoperiod, or the changing length of day and night. Most species, including the Magellanic, yellow-eyed, rockhopper, fiordland, Snares, erect crested, macaroni, royal, gentoo, Adélie, chinstrap, king, and emperor penguin, are in this category. The African, Humboldt, and little penguins sort of fit with this group, but they have a wide latitudinal range, and the northerly colonies are affected by sea surface temperature and other factors as well as photoperiod. The Galápagos penguin, which in its equatorial location has little change in photoperiod, is subject to environmental swings of sea surface temperatures and currents, the most extreme of which is El Niño, a warm-water intrusion from the western Pacific. As the sea surface temperatures get warm, the cold water that normally bathes the coastal areas is forced to deeper depths. With this change, food becomes scarce and breeding ceases, or it fails if reproduction has already begun. The Humboldt

Table 6.1. Approximate laying dates and duration of incubation and fledging.

Species	Laying	Incubation (days)	Fledging (days)
Little penguin	Aug. to Oct.	33	55
Adélie penguin	Oct. to Nov.	37	50–60
African penguin	June and Nov.	38	73–96
Chinstrap penguin	late Nov.	35	54
Emperor penguin	May to July	64	150
Erect crested penguin	Oct.	35	ca. 70
Fiordland crested penguin	July	32	75
Galápagos penguin	any month	40	60
Gentoo penguin	June to Nov.	37	80–100
Humboldt penguin	any month	41	70–90
King penguin	Nov. and late Feb.	54	300–390
Macaroni penguin	Nov.	36	70
Magellanic penguin	Oct.	40	60–120
Rockhopper penguin	Sept. to Nov.	33	70
Royal penguin	Oct.	35	65
Snares penguin	Sept. to Oct.	33	ca. 90
Yellow-eyed penguin	Sept. to Oct.	44	106

penguin is affected, to a lesser extent, as well. But these species will breed anytime conditions are good, and sometimes this means they breed more than once a year.

The king penguin has the most extended breeding cycle, but it is regular and synchronous. It cannot fledge a chick (bring it to the stage of development where it will voluntarily leave the colony) in a single season but must nurture it through the winter, or for about 14 months. This schedule puts the birds out of synchrony with the annual cycle. If eggs are laid in November, the parent penguins will not be ready to breed again until more than a year later, when they will lay in February. Thus, breeding begins either in the spring for early breeders or in midsummer for late breeders. The chick fasts during the long winter absence of the adults. Unfortunately, the chicks hatched in the autumn, meaning April, do not grow large enough to survive the winter fast, and death is almost certain for that cohort of late breeders.

In general, all penguins except possibly emperor penguins return to the same colony and sometimes even to the same nest site each year. There are advantages to this routine: it helps the members of the pair find each other again, and it means they are familiar with the area, their neighbors, and the route from the sea to the nest site.

Rockhopper penguin with a dying chick in the foreground and a healthy chick leaning against the parent.

A gentoo penguin colony on Couverville Island, Antarctic Peninsula. The ice shows stains of red guano from the krill that make up the penguins' diet.

Rockhopper penguin male and female, showing the difference in beak size. The female is in the foreground. The difference is so slight that determining the gender of a lone individual would be difficult.

Emperor penguin pair with chick.

Adélie penguins leaving on a trip.

Chinstrap penguin stained with krill and squid.

Concentrated salt gland solution dripping off the beak of an Adélie penguin.

Emperor penguin feeding chick.

Group of emperor penguins returning to the ice edge from the colony.

Black-browed albatross fighting and stealing material from a rockhopper penguin's nest. The pilfering was not without a cost; note the bloodstains on the rock, bill, and head of the albatross.

Beak dueling between two royal penguins.

Curious gentoo penguins at a photographer's tripod in the Falkland Islands.

Rockhopper penguins at the colony.

Yellow-eyed penguin on nest with chicks.

Ecotourist at the edge of an emperor penguin colony. Sit down and wait, and the penguins will come.

Emperor penguins in typical linear formation as they walk across sea ice.

Chinstrap penguin group nesting; another example of the spacing maintained by the territorial pairs.

Do penguins nest only one time a year?

In most cases penguins breed only once a year, because of the seasonal limitations for raising a chick or chicks. However, if the location has relatively mild seasonal changes, and if the time from laying to fledging the chicks is short, some species may raise more than one brood per year. Such mild conditions often occur for the little penguin, and this allows them, in a good year, to breed a second time. The same may be true for the equatorial Galápagos penguin and, to a lesser extent, for the more widely distributed Humboldt penguin, both of which may breed any time of the year and more than once in the same year. For both the little penguin and the Galápagos penguin, the long mild season is enhanced by their small body mass, making it more likely that they can breed multiple times in a year. With the small size usually comes the advantage of a shorter nurturing period, and when food is abundant, the chick hatchling grows more quickly to fledging size, as well. All other species are committed to a single brood, although some may breed again if the first clutch fails early. This is impossible for the emperor penguin, breeding on annual sea ice, where there would be no time to raise a chick from a delayed laying before the fast ice would break up. This short season requires a race against time for the breeding pair, and the fledgling has only half the body mass of the adult. No other penguin fledgling begins independence at so low a body mass relative to its parents.

Gentoo penguin with a chick near the time of fledging, which is nearly as large as the adult. In contrast, a fledged emperor penguin chick has half the body mass of the adult.

An additional problem for the emperor penguin male, if failure occurred during incubation, would be finding a mate, because all the females (unlike those of other species, which share the incubation chores) are away foraging for the entire two-month incubation cycle.

How many eggs do penguins lay?

Fifteen of the 17 species of penguins lay two eggs within a short time, and rarely some will lay a third egg if one of the first two is lost (the gentoo and Adélie penguins may do this). King and emperor penguins lay a single egg, which is supported off the ice or sand on the foot of the parent. Functionally, the *Eudyptes* (crested penguins) lay a single egg because the first egg is almost always lost early during incubation. In this group of penguins, the first egg is smaller than the second egg. Macaroni penguins have the greatest disparity between the two eggs; the first egg may be only 60% of the mass of the second egg. Most mortality is of the first egg during incubation, and in the macaroni only 3% of the first eggs hatch. The interval between the laying of the two eggs is about four days, and often the first egg is lost before the second is laid. The size of the egg of all penguins ranges from 450 g for the emperor penguin to 52 g for the little penguin. The eggs of penguins are smaller relative to their body mass than in most other bird species. For little penguins, if the first clutch of eggs is lost, the penguin will lay another clutch, but double clutching in other species is rare to nonexistent.

Macaroni penguin size disparity of eggs.

Are all hatchlings in a penguin nest full siblings?

Occasionally a pair will take over and incubate an egg that has rolled out of another nest. This is rare, as is infidelity in a pair, and almost all eggs within the nest are fertilized by the same male.

Are all penguins in a colony related to one another?

In general it is thought that little immigration or emigration occurs in a colony. This conclusion is based on studies of banded birds. It is usually assumed that banded birds that do not return to the colony did not survive after leaving the colony. However, searches for banded birds at other colonies are infrequent. In recent years DNA studies have shown some evidence of the relatedness of birds in a colony and in neighboring colonies. Genetic exchange is small, and currently the strongest evidence is that penguins are faithful to their natal colony and hence are related to some degree. The larger the colony, the more distantly related some subcolonies, groups, or suburbs may be to more distant suburbs. The term *suburbs* is not frequently used for subgroups within a colony, but I prefer it: in a city that becomes too dense, there is a migration to areas nearby, and the same thing happens with emperor penguins if they have the luxury of a large snowfield on the boundaries of the original colony. In nest builders, the movement to a suburb occurs as soon as the birds arrive and seek out a suitable nest site. In the nonnesting king penguin, which is territorial, the late arrivals have no choice but to establish an incubation site on the periphery rather than in the core of the colony. Emperor penguins collect together during incubation and, if conditions allow, will disperse from the core of the colony as the spring temperatures rise and there is no further need for huddling. In regard to genetic relationships, the further displaced one suburb is from another, the more likely that a genetic difference exists. The new recruits are forced to the edges of the main colony or to new subcolonies.

Adélie penguin colony. The colonies of this species range from a few hundred birds to more than 100,000 pairs.

How is the sex of a penguin determined?

In two words, "not easily." There is little sexual dimorphism (obvious difference) in penguins, as in many other birds that do not rely on attractive displays of the feathers by the male. Male and female look alike, but for the discerning observer there may be subtle differences in body and bill size and coloration of the auricular patches or crests. In contrast to birds of prey, in which the female's body mass is distinctly larger than the male's, in penguins the male is larger than the female. This is most obvious in the male emperor penguin at the beginning of the courtship fast. The fast for the male will last about four months, and that of the female extends only through the courtship, breeding, and egg-laying cycle. After that two-month interlude, she is able to break her fast and depart from the colony for her two-month foraging trip.

The period of copulation is the time of opportunity to determine the sex of a pair by simply observing which one is on top, marking the birds as male and female. If breeding occurs in a muddy area, then the birds with footprints on their backs are females. Another way of differentiating between males and females is by their different greeting calls. If it is not possible to discern the sex of the species with confidence by visual or acoustic means, it becomes a more technical and expensive process. The males and

King penguin copulation with assistance or interference.

females can be distinguished by using a cloacal scope to basically look into the bird's cloaca. This procedure is technically difficult and is traumatic for the bird. One of the easiest means of determining gender is DNA-based analysis of a penguin's feathers. However, this procedure requires sending the feather to a special lab. It is expensive, and there is a time delay, often of months, before the result is received.

How long do penguins care for their young?

The time from hatching to fledging varies depending upon the size of the penguin and, in one case, the special circumstance of raising the chick. The growth rate of the king penguin is too slow for chicks to be developed enough to fledge before winter. In this case the chick is fattened to a level at which it weighs more than the adult and then endures a winter fast of several weeks while the adults are on extended foraging trips to the edge of the pack ice. This is the longest fast of any bird chick. The total time to fledging is the longest in this species, taking 10 to 14 months. The delay in fledging is because of the fast. King penguins raised from the egg at Sea World molt their down and fledge at five months when fed continuously after hatching. This artificially enhanced nurturing period is similar in a way to the natural process of the emperor penguin, which has the longest uninterrupted development of any penguin before it fledges. The shortest period of development after hatching and before fledging occurs in the little, Galápagos, chinstrap, and Adélie penguins, and it is a little less than two months. For all other penguins, the hatching-to-fledging period lasts from two to 14 months.

Fat king penguin chick just before winter fast. At this stage the chick weighs more than the adult and is prepared to endure one of the longest fasts of any immature bird or mammal.

How fast do penguins grow?

The growth rate of the chicks of all penguins is a logistic curve: the initial rate is slow, and it is followed by an increasingly fast rate throughout most of the development; then the rate slows down toward the end of nurturing before the chicks fledge. Overall, the average rate of growth ranges from 80–90 grams per day for the king and emperor penguins to 15 grams per day for the little penguin. The macaroni penguin, one of the intermediate-sized penguins, grows at about 45 grams per day. The qualifiers in these growth rates are these:

1. The emperor penguin takes its chick to fledging at about 12 kg in five months, and because the onset of winter presses this species more than any of the others, the chicks are only half the mass of the adults at this stage. They are covered in 60% down when they go to sea.
2. The king penguin growth rate stalls after four months with the onset of winter. Through the winter the chick loses about half its body mass before resuming regular feedings the following spring. By the time of fledging, chicks are almost the same mass as the adults (9 kg), and they have fully molted their down.
3. The little penguin growth rate is the slowest. This is for each of the two chicks until they are fledged, at which time they have molted their

Burrow-nesting macaroni penguin parent with two midterm chicks.

down. Their 800 g weight is slightly smaller than the adult body mass of 1 k.

4. The single remaining macaroni chick weighs about 3.2 kg at fledging and has only 80% of the average adult body mass at this time in the cycle.

It should also be noted that all of these mass changes are highly variable, depending on the foraging conditions in any given year. Related to body mass variation is the fledging date, which naturally varies depending on foraging success during the season. I have noted that the date of the first emperor penguin fledgling to arrive at the ice edge can vary as much as two weeks at Cape Washington. The date for the last of those that are going to leave is about three weeks later.

How can you tell the age of a penguin?

The age of penguins, like that of most if not all birds, cannot be determined accurately. Mammals' ages can be measured by counting the growth layers in the dentine or cementum of the teeth, but in birds, which do not have teeth, there is no such convenient structure. The best data on the age of penguins come from records of banded birds. Since penguins are long-lived, they often outlive the investigator, or at least the research projects that need to include the age of birds. Molecular procedures offer the most hope for learning their age. Telomeres are the end caps of linear chromosomes and protect the chromosome from DNA loss during cell division. During the process there is loss of telomere length, and this is related to aging. According to some investigators, it may be possible to mark the age of some long-lived birds by measuring the degree of shortening of telomere restriction fragments in some tissues.

Newly hatched gentoo penguin chick.

Gentoo penguin adult with two chicks following closely in hope of being fed.

How long do penguins live?

The fact that penguins are potentially long-lived birds means that it is almost impossible to verify the age of the oldest birds. Most species probably have the capacity to live 20 years, and, indeed, a French investigator has estimated that the mean longevity of emperor penguins is 20 years. Calculations also indicate that 1% of emperor penguin hatchlings may reach 50 years of age. This may not be so far-fetched considering that some

Sea World emperor penguins captured as subadults or adults in 1976 were still alive and well in 2012. Beyond the calculation from some derived table of banded birds, the determination of longevity in wild penguins is almost impossible. The oldest known little penguin was 26 years old and still alive and wild. A banded African penguin was reported to be 27 years old. The oldest known captive penguin is a male Humboldt penguin at the Brookfield Zoo who was 43 years old as of 2012. A close second is the 42-year-old male Humboldt penguin at Sea World.

Chapter 7

Food and Feeding

What do penguins eat?

The 17 species of penguins occupy different ecological niches, or places within their ecosystem. Penguins live in widely different habitats, from the tropics to the south polar seas. Despite these differences, they are consistently similar in their food habits. Most feed at shallow depths of less than 100 m, and the main food of most species is fish. Depending on abundance and season, the prey may vary. Many northern species, such as the banded penguins (*Spheniscus*), the little penguin (*Eudyptula*), and the yellow-eyed penguin (*Megadyptes*), live in the neritic zone (near shore) and feed on small fish such as anchovy or smelt. However, if other types of prey are abundant, such as squid or krill, they will probably take these foods as well. In the crested penguins (*Eudyptes*), some of the species, such as the rockhoppers, may eat predominantly small ocean fish. The fish of choice are probably lantern fish, one of the most abundant and widely distributed of all fish. Other species of this group, such as the macaroni penguin, feed on *Euphausia*, a crustacean commonly known as krill. Great swarms of these crustaceans occur in the subpolar and south polar seas, and they are responsible for the macaroni penguin being the most abundant of all penguins. The brush-tailed (*Pygoscelis*) penguins all feed on krill as well, and in two cases, the chinstrap and the Adélie penguins, they are extremely abundant penguin species. The least abundant of the brush-tailed penguins is the gentoo penguin, which is a fish and krill eater.

Perhaps the most specialized feeder of all penguins is the king penguin, which feeds almost exclusively on lantern fish. Lantern fish, in the family Myctophidae, dwell in the deep scatter zone, the layer of living organisms.

Emperor penguins going to sea in the typical hard-charging toboggan dash.

This layer was originally observed with depth sounders that caused an echo suggesting the bottom. Later it was found that the layer migrated from 100 to 500 m depths during the day to near the surface at night. Curiously, the king penguin feeds on these fish during the daytime, when it must dive to depth to find the prey. Incidentally, king penguin feeding habits are probably more thoroughly studied than those of any other penguin. Emperor penguins have an even greater diving capacity than king penguins, but their diet varies more, ranging through squid, fish, and krill, depending upon the time of year and their location. Emperor penguins feeding over the shelf in the western Ross Sea feed almost exclusively on Antarctic silver fish, a herring-sized fish. Off the shelf, over deep water in the eastern Ross Sea and Weddell Sea, their food is mainly krill. To my knowledge, most prey of all penguins is captured at midwater to near-surface depths, and few bottom fish or other prey from the ocean bottom are eaten.

Do penguins chew their food?

Because penguins, like all other birds, have no teeth, they are unable to chew. They either swallow their food whole or, in the unusual circumstance that they catch a fish too large to swallow, they may dismember it. While I was hand-feeding exceptionally large New Zealand herrings to some emperor penguins, I was impressed to see how wide they can spread their lower jaw, or mandible, to work the fish down their throats. Also, most if not all penguins have serrated projections on the tongue, which point directionally down the throat and help to prevent prey from reversing its direction and escaping back out of the beak. In almost all cases, and especially with larger fish, the prey is manipulated so it is swallowed headfirst. In this way penguins take advantage of the streamlining of the prey to force it down the esophagus, as well as flattening the fins against the body as it passes to the stomach.

How do penguins find food?

Ever since the first penguins were discovered, this subject has perplexed penguin biologists. For those that feed near shore, like the little and Galápagos penguins, the problem does not seem so complex. The penguins may use shoreline profiles and bottom features that provide clues as to where they are and where their favorite feeding areas are. However, most penguins feed offshore, or out in the ocean, which means they hunt at distances of up to hundreds of kilometers from shore. Several species have been tracked with geolocation, GPS, or satellite transmitters attached to them, to determine where they go. Penguins probably seek favored areas of upwellings, marginal ice zones, and oceanic fronts where currents converge with water of different temperature and salinity. These are areas where prey tends to concentrate, and, depending on prey type and abundance, these patches are attractive to various predators. How marine predators, including penguins, navigate to these areas is a complex problem not well understood.

Once penguins reach an area where prey is usually found, there is the issue of specifically where and how they find prey. It could be that they detect the swimming and breathing noises of other predators, such as whales, seals, or other penguins. Penguins commonly call offshore, and they may do so more frequently when feeding. Although most birds are not noted for their sense of smell, vultures, albatross, and petrels have acute olfactory sensitivity, and perhaps penguins can smell near-surface prey such as krill or schooling fish. There is some evidence that two Humboldt penguins tracked toward a concentration of phytoplankton upwind from their position. It is believed that ultimately, once in the region where their prey occurs, penguins find the prey visually during their underwater forays. This is true even for the deep-diving emperor penguin, which may descend to very dark depths. Evidence shows that penguins rarely, if ever, dive after dark.

Species that remain near shore come ashore every evening. Those that are well offshore probably stop foraging at night. Those in polar regions migrate north to latitudes where there are at least a few hours of daylight during the winter, and they enter the water soon after morning civil twilight (when the sun is just below the horizon) and continue hunting until civil

Gentoo penguin feeding 60% grown chicks.

twilight occurs in the evening. In the extreme cases of king and emperor penguins, which dive as deep as 200 to more than 400 m to catch prey, it is unknown how they find a concentration of prey at such depths. Perhaps they simply conduct random searches in areas where prey is exceptionally abundant and the chances of finding a patch of prey are good, especially for an animal that has exceptionally good night vision. I have often thought of emperor penguins as "marine owls," because their night vision must be so good when diving to depths beyond 300 m. This depth is about the limit that sunlight from the surface reaches. At this depth and beyond, visual hunters like emperor penguins must rely on bioluminescence produced by the prey or other sources in the water column. Not a bad option, considering that more than 90% of life at depth has bioluminescent capabilities. In sum, the mysteries of long-range navigation in an apparently uniform sea, the specific targeting of prey once likely regions are reached, and the final detecting of the prey during a dive are not well understood.

Are any penguins scavengers?

It is interesting to consider the probable origin of the word *scavenger*, which once referred to someone who picks up dirt or refuse. It has a special meaning in ecology for animals that feed on carrion or garbage. A wide range of animals, such as vultures, are specialists in carrion feeding. In respect to feeding on carrion, almost any animal will take advantage of refuse or carrion when it is available or during lean times. Because of varying

Skua scavenging from gentooo feeding session.

King penguin returning from a foraging trip.

circumstances, humans, grizzly bears, and other animals at some times will feed on carrion. Many marine predators may be an exception, and especially penguins. Captive penguins are taught to eat dead fish, usually of a species that is not in their normal diet. At first they are hand fed, but in time they will eat dead fish from a tray. However, I suspect that in the wild, penguins rarely if ever eat dead animals. First, dead fish or krill are rarely available at sea, and secondly, such dead animals are ignored when they are present on land and outside the penguins' normal feeding habitat. And for that, the real scavengers in polar and subpolar environments, sheathbills and skuas, are grateful. Scavenging in penguin colonies is a profitable business, as scavengers prowl about waiting to grab food that is dropped while being transferred from parent to chick.

Do penguins ever store their food?

Since penguins eat a fresh food diet, no, they do not store food, at least not in the conventional sense of a food cache tucked away on land, under some rock at the bottom of the sea, in the snow, or under the sea ice (in the case of buoyant food). Unconventionally, it has been demonstrated that during the prolonged foraging trips of king penguins, they have a way of reducing the acidity and activity of the stomach so that some of the fish they catch are not digested rapidly. As a result they are able to travel up to a few hundred kilometers over the course of a few days and return to their chick to feed it this "store" of food. Even more remarkable is the ability of the incubating king penguin, in the last stage of incubation, to preserve food in the stomach with spheniscin, a recently discovered antimicrobial peptide. If the female does not return on time, the father still has food in his stomach to pass to the newly hatched chick. Whether this delayed digestion occurs in other species of penguins has yet to be documented. The most likely candidates are those with long incubation times or those species that go on extended foraging trips while nurturing chicks, such as the macaroni, royal, and emperor penguins (see chapter 5).

Chapter 8

Penguins and Humans

Do people ever have pet penguins?

In most countries, if not all, it is illegal to sell or trade penguins without special permits. If someone were able to overcome this obstacle, the person would find that penguins make pleasant pets. However, their needs are great, as described in the next section. If those details are not discouragement enough, then, yes, some species do make pleasant pets, companions, or neighbors. Usually, a penguin that has turned into a pet is one that was found stranded, often well out of its range, and in need of rehabilitation. The most famous and most recent case is that of an emperor penguin juvenile that was found ashore at Peka Peka beach, New Zealand, on June 20, 2011. It was held at the Wellington Zoo and given expert veterinarian care, but it was not a pet and was kept isolated as much as possible from human contact. Then it was put back on a research ship heading to the Southern Ocean, where it was released on September 4 with a satellite transmitter attached to keep track of its whereabouts after returning to sea. Sadly, after four days of tracking, the signal ended, and there were no further signals received.

There are several other examples of penguins as pets, but in general the relationship does not work out. In one case, the African penguins are not pets but neighbors in the small community of Boulders, near Cape Town, South Africa. Some of the residents are pleased to have the birds, even if they are not pleased to see the numerous tourists that come to see the birds. Other residents are not happy and would prefer to see the birds leave, which they are not likely to do. They are a well-established colony on the mainland of Africa, and it is growing—one of the few African pen-

African penguins at Boulders, South Africa.

guin colonies that is growing. Their success has much to do with the lack of land predators because they are living among tolerant humans who keep the predators away from the area.

There is a king penguin that lives with a family in Japan. The YouTube video of it has received more than 6 million hits. It is stated that the bird is over 10 years old, but no information is given on the source of the bird. The video follows it on a walk in a dense city area, where it goes to the fish market and is fed a large fish. A good demonstration of how large a dead fish penguins can swallow is shown as it gets a reward for picking up another fish and returning home. Home looks like a walk-in refrigerator. No mention is made of how long the family has had the penguin.

How do you take care of a pet penguin?

Despite the impression on the YouTube, penguins are not dry-land birds. If you have a lot of money to devote to buying fresh or frozen fish, are prepared to feed a penguin about 10% of its body mass per day, and you have a large, clean, saltwater pool with cold water for it to swim in, then you have the basic needs. In addition, penguins' hygiene is important, and they produce an impressive amount of guano, which would need to be disposed of daily. If you are a gardener, then the guano is a bonus as an excellent fertilizer. Penguins will quickly learn to be fed by hand, and soon afterward they will pick up fish from a tray. Overall, penguins may be the most care-intensive pet you could ever have, and one of the least likely to show any affection or gratitude for all your efforts.

Once I was part of a group that held emperor penguins captive for several weeks and trained them to hand feeding. The birds would follow us around the corral, greet us at the gate when we arrived to clean the corral, and tolerate mild kinds of handling. There were about 10 animals, and they all became recognizable to us as individuals, based on several behavioral traits. Yet, at the end of their captivity, when we opened the corral gate at Penguin Ranch, they left as a group, with much trumpeting to each other, and never looked back. They had bided their time as prisoners and made the most of their captivity, including gaining weight. Then they left with a dignified walk to the ice edge. In my view, they had never been happy collaborators, but they tolerated the situation without frantic pacing or attempts to push down the fence and escape.

Are penguins dangerous?

No, they are not dangerous to humans. Most live in remote areas, with little exposure to humans, and do not exhibit the flight response of most birds or mammals. This behavior leads to a misinterpretation that they are friendly; in fact, they are curious but not friendly. Their apparent friendliness ranges from curiosity to indifference. If they are handled against their will, there is a possibility of a painful wing whack, resulting in bruised arms or a bloody nose. At the worst, an eye could be lost from a well-directed bill poke, although I have never heard of such an accident.

Do penguins feel pain?

There are several types of pain, but here we are most interested in peripheral, or nociceptive, pain, and if we use the most basic definition of such an unpleasant stimulus, penguins are like all vertebrates and many lower forms of life in that they do respond to this kind of stimulus. It is important to have a response to pain in order to avoid the many bad experiences within the environment. They can range from unpleasant to life-threatening. Examples of such pain for penguins are burns, as when penguins walk through hot-water runoff from thermal streams at Deception Island, Antarctica, and crushing or tearing pain during a territorial fight or when caught by a predator. The most appalling sight of a dead penguin for me occurred when I came across one that had been sleeping under a ledge of ice and the 25 kg chunk had broken off and landed on the hapless bird. From the accident scene, it was clear that ice had pinned the bird but not killed it. Over a period of certainly days and more likely weeks, the bird had starved to death.

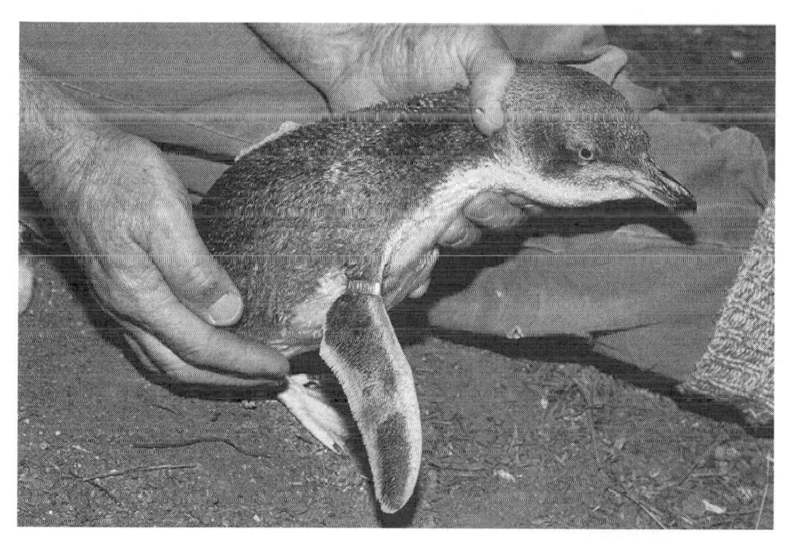

A fighting and banded little penguin that damaged the handler's thumbs.

Leopard seal with a freshly caught Adélie penguin. The bird will be flayed before consumption.

What should I do if I find an injured penguin?

In most of the world, that is a very, very unlikely prospect. For people who live near the seashore in the southern part of the Southern Hemisphere, there is a possibility of finding a penguin injured and stranded on a beach. The most likely locations for this to happen are in the area of Boulders, near Cape Town, South Africa; the southern shores of Australia; South Island of New Zealand; and along the shores of Peru, Chile, and Argentina. Unfortunately, starving Magellanic penguins even come ashore in Brazil. What you do is a matter of personal ethics. Do you assist the bird and bear the burden of an expensive veterinary bill, or do you walk on by?

Such an event can occur in the Cape of Good Hope area, where oiled penguins are found from time to time as a result of oil spills from wrecked tankers or from workers cleaning their tanks offshore. In any of the regions mentioned, if the bird is clearly sick, the best policy is to put it in a container to keep it away from local predators such as dogs and cats and to call a local aquarium or veterinarian. In addition to African penguins, Magellanic and little penguins may have similar unfortunate experiences in the areas where they live. These are probably the most commonly beached penguins, but the most newsworthy are emperor penguins. As mentioned in regard to the recent stranding in New Zealand (see "Do people ever have pet penguins?"), if a member of this species is discovered there, it is far out of its range and large crowds will gather. Over the years, young emperors have turned up in South Australia and in Tierra del Fuego.

What should I do if I find a penguin crossing the road?

For those few people in the world who live near penguin colonies, such a question of responsibility may arise. If you actually see a penguin crossing the road, leave it alone, unless it is entering a clearly dangerous area. If the road is a long way from the seashore or a colony, then it might be a good idea to catch the penguin and return it to the seashore. Otherwise, just act as a crossing guard until it is safely across the road.

What should I do if I find a penguin laying eggs?

Definitely leave it undisturbed and alert the property owners of how lucky they are. They no doubt are already aware of the breeding birds, because penguins have some rather loud calls that are made routinely while preparing to mate, selecting a nesting site, and laying the eggs.

What should I do if I find a baby penguin?

Leave it alone. The nest is probably not far away. This will be easy to determine in a colonial species such as the African penguin, but not so easy if it is a yellow-eyed penguin, because yellow-eyed penguins tend to hide their nests in the brush. There may be a fine for disturbing or moving the bird.

How can I see penguins in the wild?

One of the best places to see penguins in the wild is at the sanctuary for little penguins on Phillip Island, South Australia. More than a million visitors a year go there and pay to see penguins arrive at the colony in the

Magellanic penguin at the edge of the very large colony at Punta Tombo, Argentina.

evening when they come ashore and feed their chicks. There are also other places to see little penguins, especially in New Zealand, near Christchurch or Dunedin, where you can rely on a guide or tour the area on your own. Other areas where self-guided tours are possible are at Boulders, South Africa, where African penguins have a colony in a picturesque setting near Capetown. There is a small colony of Magellanics within a 30-minute drive from Punta Arenas, Chile, and a large and famous colony in Punta Tombo, Chubut Province, Argentina, with bus transportation. This colony is about 1,000 miles south of Buenos Aires—not an easy trip, but well worth the trouble if stops are made at Golfo Valdez as well to see southern right whales, killer whales, and southern sea lions. Shorter but also extended overland trips may be made to see Humboldt penguins in northern Chile or southern Peru, for instance at the famous Punta San Juan wildlife sanctuary. The place that is most famous of all is also the most difficult, because of its location offshore from Ecuador; that is the Galápagos Islands, where the lucky visitor might see Galápagos penguins. As for seeing sub-Antarctic or Antarctic penguins in the wild, you will need time and money. There are no self-guided tours, but the wildlife experience of seeing a penguin colony will be one of the best in your life, even if you live to be a hundred.

Do people feed wild penguins?

No attempt should be made to feed penguins in the wild. Penguins would never approach a person to be fed, but their curiosity may be misinterpreted as that of a freeloader. It is unlikely that you would have the correct food, and trying to approach a penguin to feed it would disturb the bird and others in the colony. In short, penguins are not fed in the wild by humans.

Chapter 9

Penguin Problems (from a human viewpoint)

Are penguins ever pests?

It is difficult to imagine penguins as pests, especially if you are a penguin lover such as I am. However, if you live near an area where penguins breed, the calls they make to their mates and the presence of their nests on your property may become an annoyance. There are only a few such places, and the rarity of such close and peaceful associations with humans makes them world treasures. Perhaps the most intrusive of all such potential nuisances occur in the community of Simon's Town (a mainland site near Cape Town), where African penguins have established a colony at Boulders Beach. Instead of making burrows on the beach, some of the birds resort to establishing nests in the yards or under the porches of their nearby human neighbors. Even in this community, I suspect that the greatest annoyance comes from the tourist traffic drawn by the penguins rather than from the penguins themselves. It would be interesting to know the effect of penguin presence on the property value of the homes near the beach at Boulders. Perhaps it raises the value by 10% to have an active penguin nest in the garden.

Do penguins reduce the number of fish or krill?

It is uncertain how strong an effect penguins have on fish or krill populations; probably no virgin fishery that is exploited by the birds has ever been studied. This question has been asked time and again in regard to penguins, other seabirds, and marine mammals. The concern comes from fishers, who do not ask the question until they also have been exploiting the

Braying African penguins; they were once called jackass penguins.

fish. It is prompted because in all cases the fish population is declining, and the industry is looking for a scapegoat. A telling aspect of this is that fishers often select areas where there was a thriving population of fish and marine predators, such as penguins, before they themselves began fishing. In many cases the decline in the fish is a result of overfishing, not predation by the marine predators, which have been in balance with the fish population for decades.

A classic example is the anchoveta population off the coast of Peru. This population of fish experiences wide swings in abundance. The swings were formerly driven by the effects of the cyclical El Niño ("the Christ Child"), so named because it occurs around Christmas. Although addressed most often locally to Peruvian waters, El Niño seems to have a worldwide effect. In brief, for the Pacific there is a reversal or weakening of surface air pressures between the eastern and western tropical Pacific. Normally winds blow from the Andes Mountains in the east across the tropical Pacific to the west. Near the shore this results in a coastal upwelling that results in a shallow thermocline, producing some of the most fertile waters in the world. (A thermocline is a boundary layer where the water abruptly changes temperature at a certain depth and then more slowly changes temperature as the depth increases.) The anchoveta thrive and so do the marine predators along the coast. At one time (in about 1954), before the local fishery became so industrialized, the population of the guanay, a species of cormorant, was more than 20 million. When the winds reverse and the thermocline deepens, the anchoveta remain at depth. Then the guanay starve and fail in breeding, and the population crashes along with that of the Humboldt and Galápagos penguins, the Peruvian booby, and the Chilean brown pelican. The population of guanay is now estimated to be only

Group of Humboldt penguins.

about 3 million. It never recovered because the fish population never was allowed to recover to preexploitation levels. Even during the El Niño, fishing continued.

A similar example exists for the African penguin, which in the early 1900s had a population of over a million adult birds. Over the course of the past 100 years, fishing efforts of humans have increased and are in direct competition with the penguins for the same species of fish. The present population of penguins is about 20,000 to 30,000. However, the penguin population decrease is not all a result of competition with the fishery; there is extensive offshore oil pollution from tankers rounding the Cape of Good Hope, and periodically hundreds to thousands of oiled penguins come ashore. Owing to the efforts of the Southern African Foundation for the Conservation of Coastal Birds, in Cape Town, many of these birds are rehabilitated and released back into the wild. Because of these various insults to the penguin population, it continues in a steep decline, and this penguin has recently been listed as endangered by both the International Union for the Conservation of Nature and the U.S. Fish and Wildlife Service. The U.S. Fish and Wildlife Service is very conservative about endangered species; this is only the second species of penguin on its endangered-species list.

In summary, the reduction of prey population by penguins is uncertain. Intuitively, one would conclude that thousands of penguins eating fish must have a negative effect on the fish. However, both the fish and the penguin populations may be sustainable in the interaction that is maintained. In most cases the birds may be taking mainly the young of the year, so that the most reproductive element of the fish population is not strongly affected. It has been shown recently by a group of biomathematicians that the most sustainable exploitation, allowing stabilization of fish populations, is to take the younger fish. This is not what industrial fishery operations usually do.

Do penguins kill endangered species?

Because the penguin is an intermediate predator (meso apex predator) in the food chain that feeds in the nearshore to midocean zone that depends on abundant prey, it is not likely that penguins will kill many, if any, endangered fish or squid species. In fact, most marine species that are listed as endangered, such as the great white shark, are too big for penguins to eat. It is more probable that penguins will be eaten by endangered species than the other way around.

Do penguins have diseases, and are they contagious?

It is likely that every species of bird, and for that matter all vertebrates, have diseases and that many are contagious. In regard to penguins, this is not an especially popular research topic. A review of mostly New Zealand penguins showed that there were eight types of disease involving about nine different species of penguins. The viral diseases for which evidence has been found in penguins are the following:

1. *Arboviruses, or viruses transmitted by arthropods.* Antibodies against the *Flavivirus*, which can cause encephalitis and west Nile virus, among other diseases, have been found in rockhopper, royal, king, and Adélie penguins.
2. *Orthomyxoviruses, or influenza viruses, a strain of which is found in birds.* As for penguins, antibodies were present in some Adélie penguins.
3. *Birnaviruses, which can cause bursal disease in birds.* The bursa of Fabricius is a specialized organ that is important for the immune system. Without it, antibody production ceases, and the young bird is susceptible to a variety of diseases. It can cause much mortality in poultry. Antibodies against infectious bursal disease were detected in emperor and Adélie penguins in Antarctica, but there was no evidence of disease.
4. *Paramyxovirus, an RNA virus that causes acute respiratory diseases.* One form is responsible for Newcastle disease in birds as well as related diseases in seals and whales (phocine distemper and cetaccan mobillivirus). Newcastle disease is of special concern because the spread of the disease in commercial poultry can cause significant monetary losses. Of course, such losses catch the attention of any disease control center and often result in the quarantining or inspection of bird products from a foreign

Group of king penguins in early spring at South Georgia Island.

area. This includes penguins, but the risks are minimal, considering that evidence of Newcastle's disease has been found in penguins only a few times, in a few species. The penguins identified are king, royal, rockhopper, and possibly Adélie.

Two bacterial diseases are of concern. The first is Avian psittacosis, or, more correctly, ornithosis in birds other than parrots. This is another disease for which antibodies have been detected in rockhopper, royal, gentoo, emperor, and Adélie penguins, but there was no evidence of the disease. The second, *Pasteurellosis*, is a bacterial species that causes avian cholera in waterfowl and was isolated in some sick rockhopper penguins on Campbell Island. Of the diseases mentioned, only this last one was expressed as an actual sickness in birds. All of these diseases are potentially contagious, but in only a few cases is there evidence of transmission of the disease to other penguins or other birds. A few mass die-offs of penguins have occurred. Two examples are the die-off of 100 gentoo chicks at Signy Island in 1969 and the 1972 die-off of 65% of the Adélie penguin chicks near Mawson Station. No causative agent was determined.

Most if not all penguins are infested with ectoparasites (exterior parasites) such as fleas, ticks, and lice. In some cases the infestations can be heavy, and I personally noted intense patches of ticks in the head feathers of king penguins at one colony in the Crozet Archipelago. Penguins also have a variety of endoparasites (interior parasites) that range from worms that occur in the intestine, liver, and gall bladder to avian malaria. The latter occurs naturally in little and African penguins and is especially dangerous for captive polar penguins held in zoos, because of the risk of contracting the disease from the local seagulls.

Rockhopper penguin on a nest in the Falkland Islands.

Do people eat penguins, and is it safe?

People do eat penguins. I wish I could say that it is unsafe, but it is not if the meat is cooked well. In the Antarctic exploratory years of the mid-1800s to the early 1900s, Antarctic penguins provided an important part of the diet for the explorers. African penguins were probably a food supplement for Bartolomeu Dias as well in 1488 when he anchored in Mossel Bay. However, it did not stop there. Some locals who live near penguin colonies still use penguins and their eggs as a supplementary food. Egging was a popular activity on St. Croix Island, South Africa, and thousands of eggs were collected until the 1950s, when it became clear that it was not a sustainable practice. In the Falkland Islands in 2010, licenses were issued allowing the collection of 2,262 gentoo penguin eggs at 20 colonies, and 2,059 eggs were collected. An unpublished report indicates that there are 120,000 breeding pairs among all of those islands.

Chapter 10

Human Problems (from a penguin's viewpoint)

What are the least and the most abundant penguins?

It is a curious fact that there is no total population estimate for the penguin that lives in closest association with humans, the little penguin. The various colony populations of this wide-ranging species are difficult to estimate because the birds are cryptic in most of their habits. They come and go from their breeding areas most frequently at night, and they nest in burrows, in caves, and under dense foliage. The most remote of all penguins, the emperor penguin, recently became the best counted of all species. It is the ideal species for counting, because it breeds on fast ice (the relatively narrow band of ice along the shoreline that does not move) in completely exposed areas, and the black backs of the penguins contrast sharply with the white environment. There is no cryptic behavior in this species; an aerial photograph shows the detailed characteristics of the colony. Recent satellite data indicate that the world breeding population of emperor penguins is about double that of earlier estimates reported by BirdLife International in 2012. (You can find details at the BirdLife International website, www.birdlife.org.) This is not surprising, because the satellite revealed several new colonies and many of the known colonies were not counted because of their remoteness. It should not be concluded that emperor penguin numbers have increased in recent years. In fact, there are not enough long-term statistics on most colonies to estimate trends.

Additional information obtained from BirdLife International is the population estimate of the least abundant species, the Galápagos penguin, which has only 1,800 birds in the wild. Significantly, this species is exposed to strong cycles of El Niño, which have had a powerful effect on the popu-

Table 10.1. Population estimates and trends of mature individuals.

Species	Total	Trend	Status
Macaroni	18,000,000	declining	vulnerable
Royal	1,700,000	stable	vulnerable
Erect crested	49,000	fast decline	endangered
Fiordland	5,000	declining	vulnerable
Snares	60,000	declining	vulnerable
Northern rockhopper[a]	300,000[b]	fast decline	endangered
Southern rockhopper	750,000[b]	stable to fast decline	vulnerable
Magellanic	4,000,000[b]	declining	near threatened
African	52,000	declining	endangered
Humboldt	ca. 8000	declining	vulnerable
Galápagos	1,800	declining	endangered
Yellow-eyed	4,800	declining	endangered
Little	unknown		least concern
Gentoo	520,000	declining/increasing	near threatened
Chinstrap	8,000,000	uncertain	least concern
Adélie	4,500,000	uncertain	least concern
King	2,000,000	increasing	least concern
Emperor	575,000[c]	uncertain	least concern

Notes: This information is from BirdLife International's 2012 International Union for the Conservation of Nature (IUCN) Red List for birds, downloaded from www.birdlife.org on April 30, 2012. The date is important, because every year the numbers tend to change, and usually to lower values.

[a]The northern rockhopper is designated as a new species based on the IUCN conclusion.

[b]The mean of a wide range.

[c]A recent population estimate from satellite imagery that results in nearly doubling the estimate from BirdLife International.

lation. In 1983 the population declined to 25% of its previous level, and it has not returned to pre-1983 levels. The macaroni penguin is the most abundant of the 17 species, at 18 million individuals. One wonders how the total was determined from the 216 colonies widely distributed on the islands around the sub-Antarctic and southern Chile. Because of recent population declines, especially at South Georgia Island, this species is considered by BirdLife International to be vulnerable.

How are penguins counted?

Not all penguin population estimates are created equal, except that they are all estimates from breeding populations. Each species presents its own set of challenges. The three major characteristics of penguin breeding habitats are remoteness, social behavior, and habitat selection. Access

Galápagos penguin calling.

is a major problem for those birds that live on remote islands. Some can be reached only by boat and the islands are so rugged that it is difficult or impossible to survey the breeding habitat by traveling over land. Antarctic breeders have the ultimate inaccessibility, and if it were not for air support and satellite coverage, it would be impossible to count any but a few colonies found on the coasts of Antarctica. Social behavior is paramount for the task of counting any species. The birds' sociality results in colonies that can be found and counted, although for many species, neither the finding nor the counting can be done without considerable effort. Habitat selection for a colony adds to the complex mix of assessing the number of breeding birds in known colonies. By nature most penguins are not cryptic, but in the lower-latitude species, nesting in caves or burrows is often necessary to escape the heat of their desert environment, as well as to gain protection from predators.

Galápagos penguins, which live at or near the equator on very hot and dry desert islands, are an example of a cryptic nester. This rarest of penguins cannot be counted directly in most places where it occurs, so estimates are accomplished indirectly by mark-and-recapture methods. Based on the recovery of birds that were previously caught and marked, and some mathematical manipulations, an estimate is derived. The open-area nesting of African penguins lends itself to more direct estimates, which can be obtained by counting all active nests in a colony. This is labor intensive, as a team of counters assessing the different colonies accomplishes all counts. The project can be time consuming if there is a large population, but it can be done for a population of tens of thousands of birds. If the population is in the millions, such as that of the most abundant species, the macaroni penguin, the task is impossible by such a conventional effort as is used for the African penguin. This is especially true because there are more than 200 colonies. Willis Island, part of an archipelago of South Georgia Island, is not only difficult to reach, but the colonies there had an estimated total count in 1986 of 5 million adult birds. It is uncertain how that estimate was obtained, but from 2000 to 2003, when a solid count was carried out from aerial photographs and an automatic counting system, the estimate of adult breeding birds was more than 1.9 million.

Ironically, it is for the most remote and inaccessible of all penguins, the emperor penguin, that the most accurate and practical means of estimating an entire breeding population may be provided. No it is not done by angels, but by satellite imagery. Thanks to emperor penguins' habit of breeding on flat sea ice, and thanks to their black backs, which all penguins have, a new technology of low earth satellites (owned by Digitalglobe and called Quickbird) has made it possible to discern black-on-white images as small as 60 cm from 450 km above the Earth. With some technical wizardry for separating guano stain from clustered birds on snow or ice, and with computer analysis of the area and density of the birds, a population estimate was derived. At about a half million birds in widely distributed colonies, this population is not going to be hand-counted. However, nearly simultaneous aerial photographs of a select few colonies, which were hand-counted, have supplemented the counts from satellite images and corroborated their accuracy.

Will such advanced technology be possible for other species in the future? Possibly, for other Antarctic species, and perhaps for some of the more remote and exposed penguin colonies of the sub-Antarctic, but it is not feasible for the species that nest under rocks, underground, hidden by vegetation, or in caves. For the continental species of African, Magellanic, and Humboldt penguins, the most cost-effective method will continue to be on-the-ground or aerial surveys. Because of the wide distribution and cryptic nature of the little penguin, its total population, even though these penguins live underfoot, so to speak, will remain a mystery.

Are any penguins endangered?

Several penguin species are endangered or threatened with extinction, depending upon whose list is considered. There is much difference between the lists prepared by the International Union for the Conservation of Nature (IUCN) and by the U.S. Fish and Wildlife Service (USFWS). The penguins that the USFWS considers endangered are the Galápagos penguin, designated in 1970, and the African penguin, listed in 2010. All the others listed as endangered by BirdLife International, the listing authority for IUCN, are considered threatened by the USFWS. The endangered species according to the IUCN include the northern rockhopper (*Eudyptes moseleyi*), the erect crested (*Eudyptes sclateri*), the yellow-eyed (*Megadyptes antipodes*), the African (*Spheniscus demersus*), and the Galápagos (*Spheniscus mendiculus*). The IUCN evaluates its list about every five years; some of the evaluators are BirdLife International and specialist groups within the IUCN. The USFWS responds to petitions, and in 2006 the nonprofit Center for Biological Diversity filed a formal petition with the U.S. government requesting that 12 species of penguins be added to the list of threatened and endangered species under the federal Endangered Species Act. The USFWS is required to respond within 90 days to any petitions, and it did. The main reason for the petition was not that some of the 12 species were in danger of extinction in the near future, but the concern that global warming was the path to extinction for all of them. Apparently the USFWS did not see that as a valid reason, even though the act states that if species are "in danger of extinction within the foreseeable future throughout all or a significant portion of its range," they should be listed. In August 2010 six species of penguins were listed as threatened: the yellow-eyed (*Megadyptes antipodes*), the white-flippered (*Eudyptula minor albosignata*), the fiordland (*Eudyptes pachyrhynchus*), and the erect crested (*Eudyptes sclateri*), all from New Zealand, as well as the Humboldt (*Spheniscus humboldti*) of Chile and Peru. In September 2010 only the African penguin was given the status of endangered, because of the precipitous decline in the population over the previous few years. I was surprised that the yellow-eyed penguin was not listed as endangered because of its small population, limited distribution, lack of gregarious nesting, habitat loss, and exposure to the invasive predators of dogs, cats, and stoats. Eventually, by any reasonable definition, all cold-temperate species, which live in close association with humans, will be endangered. Note also that of the above-mentioned species, two are still officially designated as subspecies. They are the white-flippered and the northern rockhopper.

How does tourism affect penguins?

Ask this question to representatives of the tourist industry or the tourists who have visited the Antarctic and sub-Antarctic, and most will say there is no problem. Ask the scientist working in these polar regions, and the response might be mixed. Of all the ways humans contact or interact with wildlife, polar tourists are the most benign. Even though there are, at present, about 40,000 tourists traveling to Antarctica each year, most are in their twilight years and are mellow about the experience they must have. The younger, more adventuresome set is more intrusive; but, all consid-

White-flippered penguin of New Zealand.

ered, they have little effect. No Antarctic tourist shoots or eats wildlife or builds permanent structures that displace the wildlife. The tourism industry is self-regulated by the International Association of Antarctic Tour Operators. Penguin colonies are the favorite stopover, and in most cases landings are made at designated areas. Most groups have a guide who keeps tourists on specific walks around or through the colonies. Perhaps the greatest problem is the constant passage of tourists near nesting birds, especially during the incubation and brooding phase. Some sites are visited daily by at least one ship, and hundreds of passengers traipse along the penguin paths to visit the colonies. Overall, however, the intrusions do not seem to have much effect. The greatest environmental damage occurs when a ship runs aground or sinks. Any resulting oil spills can be serious and long-lasting. One of the worst oil spills, and also one of the first, occurred after the 1989 grounding of the *Bahia Paraiso* near Palmer Station, in the Antarctic Peninsula. There were about 3,800 barrels of diesel fuel onboard. Some of this fuel began leaking soon afterward, but in 1992 most of the remaining oil was recovered. In comparison, the *Exxon Valdez* spilled a disputed estimate of from 240,000 to 750,000 barrels of crude oil into Prince William Sound, Alaska, also in 1989.

Visits to the more northern penguin colonies are regulated by each country where they occur. In the Galápagos Islands, contact or likely contact with any wildlife is highly regulated, and the number of tour boats visiting the various islands is limited. At Boulders near Simon's Town, South Africa, there is a designated boardwalk for visitors to use, and a ranger stays close by. A similar situation exists for most of the colonies in South America, New Zealand, and Australia. It would be hard to find an industry less damaging to the environment than tourism, and the penguins are excellent ambassadors for promoting not only their own conservation, but that of other wildlife.

Ecotourist and king penguins, South Georgia Island.

Will penguins be affected by global warming?

Global warming or any other global change in climate patterns will affect all creatures on the planet, but the effect will occur in varying degrees and directions, depending on species and location. There is a severe warming effect under way in the Antarctic Peninsula, where three species of penguins occur. Two of these species, chinstrap and Adélie penguins, are currently at only 50% of their numbers, compared to census reports from the 1970s. The proximate cause is uncertain, but a recent hypothesis proposes that the population decrease is directly related to a drop, over the past 20 to 30 years, in the krill population, the penguins' food base. The krill population has declined substantially during this time because of the reduction in the extent of winter sea ice in this region, which is one of the fastest-warming areas of the planet. The ultimate cause of penguin mortality is correlated to the negative effects of climate change, which also results in extremes of seasonal weather. Breeding success is heavily affected by excessive snowfall in some years during the time of egg laying and brooding. In some cases nesters are buried completely while relentlessly remaining on the eggs or the chicks. Some hypothesize that colonies of emperor penguins located north of 70°S will decrease or even disappear if the earth's average troposphere temperature continues to rise and reaches 2°C above the preindustrial levels of the 1860s.

With regard to the world's penguins, we have seen the most dramatic ecosystem changes in the Antarctic Peninsula. Scientists aware of the ef-

Adélie penguins on the Antarctic Peninsula.

Adélie penguin buried as a result of a snowstorm and refusing to leave its nest.

fects on penguins urge that limits be imposed on factors leading to global climate change that are under human control.

Are penguins affected by pollution?

There are many forms of physical, chemical, and biological pollution in the marine environment. One of the most serious is oil pollution. The contamination from sinking tankers is always a surprise, and surprisingly, the largest spills do not necessarily cause the greatest mortality. The *Cas-*

tillo de Bellver, which sank in 1983, still holds the record for the sixth-largest oil spill in the world, yet it involved damage to only 800 penguins at nearby Malgas Island and Lambert's Bay. At least that is how many penguins made it to shore. Most of the oiled birds were cleaned by the Southern African Foundation for the Conservation of Coastal Seabirds, which was established in 1970. Ever since that time, there have been almost yearly events of penguins being exposed to oil, ranging from a few hundred after a collision of two tankers in 1977 to the sinking of the *Apollo Sea* in 1994. Although this cargo ship spilled only 1,300 metric tons of oil, the oil contaminated 5,000 penguins. The year 2000 was the worst for African penguins; that was when the *MV Treasure* sank near Table Bay, Cape Town. It produced a relatively small spill of 1,300 metric tons, but it endangered 41% of all African penguins. More than 38,000 birds were handled, 19,000 of which were oiled and had to be cleaned. The others, the unoiled half, were moved to Port Elizabeth and released. There may be no other region that could have handled such a large volume of birds. Since then there have been no major oils spills in the African penguin range, but before that, in every year going back to the 1970s, a few hundred or more penguins turned up oiled. Penguins seem to be a perfect sponge for oil, and some penguins are soaking up oil from unknown places and origins, which might be called "ghost slicks." Oil damages the penguin's feather coat, destroying the thermal properties of penguin feathers. It is also a poison. When grooming their oiled feathers, birds ingest many of the toxic substances embedded in their feathers. As a result the birds not only get chilled, but they become ill from serious gastrointestinal distress.

Finally, a cautionary note: no area is safe from oil pollution, even in the remotest of seas. A cargo ship, the *MS Olivia*, managed to run into Nightingale Island, of the Tristan da Cunha group, and spilled enough crude oil to coat thousands of northern rockhopper penguins, one of the five endangered species (in this book considered a subspecies) of penguins. This occurrence is rather unbelievable, since this group of islands is one of the most isolated in the world, and hardly anyone ever goes there.

Another pollutant from industry that is deadly for penguins is fishing gear, lines and nets. Some of it is lost gear that continues to catch fish, penguins, and other marine birds and mammals even when fish are no longer harvested (ghost fishing). Coastal gill nets are especially destructive to penguins. For example, 605 Humboldt penguins were drowned in gill nets off the central Chilean coast between 1991 and 1996. There are similar results from gillnetting in other areas. The most destructive are the drift gill nets, as distinguished from the bottom-attached method of gillnetting.

Do zoos and aquariums remove penguins from the wild?

Most species of penguins in captivity are able to breed within the facilities where they reside. This practice produces a supply of new birds, and some of the surplus is traded or sold to other zoos. When it is necessary to take new birds from the wild, the preferred method is to collect fertile eggs. The eggs are placed immediately into an incubator and transported to the proper quarantine facility for a required period of time. Later, after hatching and when the chicks are able to eat independently, they are moved to the viewing tanks. In this way the staff have birds of a known age, which is an important factor for the breeding program and the demographics of captive populations. In only one species, and perhaps the most sought after, is the procedure of collecting eggs instead of adult birds not possible. Since emperor penguins breed in winter, eggs are unavailable for collection. Adult or juvenile birds are collected instead. Special permits must be obtained to conform to the regulations of the Antarctic Treaty. Emperor penguins are also the least prolific of all captive species, and exchanging birds from one aquarium to another is not usually practiced.

Are any commercial products made from penguins?

At present no products are made from penguins, and no penguins are harvested commercially. It was different at the beginning of the twentieth century. At that time one company obtained a license to kill both king and royal penguins on Macquarie Island for their oil. Parts of the steam digesters used in the process to render oil from the carcasses are still standing at a place called Nuggets. This industry ceased operations in about 1920. During peak production about 150,000 birds were slaughtered per season. The marketing of oil from fossil fuels killed the penguin-oil business, which was getting heat from wildlife conservation groups, also. The present attraction and value of the birds to ecotourism probably exceeds the total value obtained from the unsustainable harvesting of the birds.

What can an ordinary citizen do to help penguins?

Ordinary citizens play an important role in the education of their fellow citizens about the importance of penguins and their habitats. Success in conservation efforts is frequently the result of activities that interested individuals promote through a variety of methods, from public lectures to fund-raising events for environmental protection programs. The emperor penguin has become an iconic symbol for the Antarctic, substantially

Steam digesters on Macquarie Island remain as a reminder of a grim episode in the history of the sub-Antarctic.

thanks to the documentary *March of the Penguins.* Other species of penguins are also becoming better known because of this attention and the resulting rise of interest in all penguins. In the past, activists have played important roles in saving penguins, and they continue to do so. Their work has ranged from publicizing the use of penguins for food and fuel to stopping a proposal to harvest Magellanic penguins for gloves and hand muffs. At this stage in the age of humankind, sometimes called the Anthropocene, we are the only species among all others on this amazing planet that can make a difference in the continued existence of most wildlife.

Are penguins important to the conservation effort?

Penguins are important to the conservation movement for several reasons, but most obvious is their role as a major symbol of wildlife, especially in the areas where they live. The African penguin is especially emblematic because of the powerful effect that offshore fisheries and oil spills have had on the population of that penguin. Both the fishing industry and the penguin population have been monitored for decades, and the correlation between the two is strong. The alarm about overfishing has been broadcast more widely because of attention to the huge decline in the African penguin population. It was more than 1 million in the early 1900s, but now it is reduced to about 50,000 adults. No wonder it has recently been listed as endangered by the U.S. Fish and Wildlife Service, a relatively conservative organization when it comes to listing endangered species. Furthermore, the predicament of the African penguin is underscored at times by news of large numbers of oiled birds coming ashore. This is an example of how important penguins can be to the broader conservation effort, because less publicity is given to many other species of animals, which are concurrently influenced negatively by overfishing and oil spills.

A clarion call for penguin awareness occurs in many other places. The situation is as serious for the Magellanic penguins, off the east coast of South America, as it is for the African penguin. Overfishing in their local waters is driving the parent birds farther offshore to hunt while nurturing chicks. Some birds are turning up on beaches of Brazil during the postmolt migration phase of their annual cycle, when they should not be coming ashore anywhere. For the past decade, there have been regular reports of starved and oiled birds being stranded on Brazilian beaches. In some cases gallant efforts have been undertaken to clean the birds and return them to more southerly beaches for release. Such heroic interventions by humans to rescue these birds are successful and laudatory. One of the positive end results is the publicity that draws attention to the possible causes of the problem. The penguins, as proxies, also demonstrate that there may be similar problems for other animals that are less visible and less inclined to come ashore.

In the Antarctic Peninsula, the influence of regional warming on penguin breeding success has been addressed in many scientific studies. One of the most recent reports about population declines in chinstrap and Adélie penguins highlights the substantial warming trend in the area, which causes a reduction in krill. Again, these species have been studied for decades and are literally more visible than other species that have not benefited from long-term population studies.

There are other examples of habitat deterioration, but the few discussed illustrate the importance of species known so well by the public. Their perilous situation becomes a powerful symbol for much of what is happening in relation to environmental conditions in their region. People relate strongly to animals as symbols of strength, beauty, and even "cuteness." As the giant panda is a symbol of worldwide conservation problems on land, the penguin is a worldwide symbol of marine conservation.

Chapter 11

Penguins in Stories and Literature

Have penguins played a role in religion and mythology?

Most religious stories were formalized before much was known about penguins. The first Europeans to see penguins were probably a part of the Bartolomeu Dias expedition, which discovered Cape Hope in 1488. Unlike Vasco da Gama, who came 10 years later, Dias cruised the entire west coast of Africa, from Cape Verde to Great Fish River, before turning back. From Port Nolloth south, Dias's group may have seen penguins at sea, but a storm blew the ships out from the near shore, and they did not make landfall until arriving at Mossel Bay, where penguins were in abundance. Proceeding from the Cape Verde Islands and well out into the South Atlantic, da Gama made landfall at St. Helena Bay in 1497 and then proceeded on to Mossel Bay. In either place his group would have seen penguins. The most detailed accounts of discoveries in this area came from the later expedition of Ferdinand Magellan, in 1519–22. The scholar and explorer Antonio Pigafetti, who wrote a detailed account of the entire expedition, probably described both the African penguin seen in 1522 and the Magellanic penguin seen in 1519, and he lived to tell about it. He was one of 18 original expeditioners who survived, from a total crew of 237.

The African penguin was the earliest-described penguin species, achieving that distinction in 1758. Appropriately, Carl Linnaeus, the originator of binomial nomenclature, described it from an account and collection done by a person named Edwards in 1747. It is fitting that the African penguin was the first penguin described, because it was the first to be discovered by Europeans, going back to the Dias voyage. Because penguins were not known widely by the developed nations in Europe and Asia until the 1800s,

African penguin, the first penguin species observed by Europeans.

it is not surprising that they have not been incorporated into tales, myths, or legends of these cultures.

What do penguins have to do with fairy tales?

Very little if anything about penguins appears in the genre of old-time fairy tales. Too little was known for them to be incorporated into the Brothers Grimm's fairy tales written in the early 1800s, or into similar works. The menagerie of animals featured in those tales mainly comprise the common wild and farm animals of that period, such as cats, rats, frogs, bears, pigs, and the big bad wolf. Penguins are a dominant character in modern fairy tales, cartoons, and animated movies.

What roles have penguins played in native cultures?

It seems that penguins played a small role. There is no fully authenticated indication of penguins in rock art or in any legends. However, coastal native cultures must have had some contact with penguins in South Africa, South America, and Australia. Perhaps interactions were limited because most species lived on uninhabited islands or barren coastal regions. Consider the locations of the Galápagos Islands, all of the remote sub-Antarctic islands, and the harsh desert regions of South America and South Africa. Reputedly there is a depiction of at least one penguin in the 2,500 images of the Twyfelfontein petroglyphs of South Africa, but evidence is scarce. Surprisingly, penguins are mentioned in reports of rock art of the Mediterranean cave of Cosquer. The second-oldest rock art in the world is in this cave. The oldest resides in the recently discovered and surveyed Chauvet

The floppy wings of the emperor penguin chick.

Cave, in Ardeche Valley, France. Those artworks are more than 30,000 years old, and the paintings are exquisite animal paintings. The earliest set of the Cosquer paintings is 27,000 years old and consists only of handprints, but the second set of animal drawings is dated at 19,000 years old. These were drawn at the end of the last ice age, and that is why this once-seashore retreat for cavemen has an entrance that is now 37 m under water. There is a 175 m inclined passage leading to an above-water entrance into a large chamber. Here are found 177 different animal figures, mostly of horses, goats, and bovines. It has been reported that nine seals and three penguins are included among the marine animals. But the three images are probably not penguins. One commentary on the paintings states that there are three great auks and one penguin. The comment should have stopped at the great auks. The idea that a penguin was represented must be a misinterpretation, although the photo of one figure easily suggests a penguin, and one would conclude that it was just that if the cave were in South Africa. At the time of the paintings, the nearest penguins would have been thousands of miles to the south, along the shores of the Namibian desert, with the great tropic belt in between. The northern ice cap at this time reached to central Europe, so the great auk likely occurred much farther south than when it met with human-induced extinction in 1844. It is plausible that the artist in this cave had seen a great auk that most likely

The Twyfelfontein petroglyphs, at a UNESCO World Heritage Site in Namibia, include a few penguins (not shown in this photograph). The rock etchings, which number more than 2,500 and date from 6,000 years ago, were made by Stone Age San hunters.

wandered into the Mediterranean during its travels in the Atlantic Ocean. Aside from these remarkable paintings, no known rock art, cave paintings, or figurines of penguins are associated with early cultures.

Of the four major areas where early man lived in South Africa, Tasmania and South Australia, South America from Tierra del Fuego and Patagonia to Peru and Chile, and New Zealand, penguin artifacts seem scarce. The Indians of southern South America were closely associated with the sea and almost certainly collected penguin eggs for food, if not the birds themselves. The Ona, the Alacaluf, and the Yaghan were small populations of hunter-gatherers who used seabirds as part of their diet, but they may have hunted terrestrial mammals and birds more extensively. These native groups were rapidly decimated soon after contact with Europeans. The last Ona died in 1999. And the languages of these South American cultures are nearly or completely extinct. The coastal Indians of Peru had large populations and were mainly farmers. Most likely they supplemented their diet with seabirds, including penguins.

Similarly, the Bushmen (San and Khoikhoi) of South Africa were hunter-gatherers and herders. The San lived mainly inland from the coast, and the Khoikhoi were mostly coastal dwellers. Their contact with penguins was probably limited, because most penguin colonies occurred (and still occur) on offshore islands, which are inaccessible to most coastal carnivores, including man.

There is little information regarding the use of penguins by southern Australian and Tasmanian aboriginals, even though aboriginals have an exceptionally long occupation in the area. The Tasmanian natives are of particular interest because they were coastal dwellers who crossed over to Tasmania during the presence of a land bridge and lived in isolation there for about 8,000 years. They disappeared after European contact in the 1700 and 1800s.

Coming full circle to New Zealand, the Maori arrived about 1,000 years ago. They made short work of the flightless moas, some of which were about 3 m tall. Because of their size and flightlessness, they were attractive prey, especially compared with the little penguin and other species of penguins that lived around the southern coasts. One seabird that was and still is harvested is the "muttonbird" or sooty shearwater. Like penguins, it does not seem to be represented much in Maori art. In short, penguins do not seem to have played much of a role in native diet or cultures, at least to the point of being represented in native art forms.

What roles do penguins play in popular culture?

Penguins have played a large role in popular culture and continue to do so today. Penguins serve as the mascot for hockey—and also for classic books. The range of interest covers the gamut from comic books to full-feature movies and everything in between, including toys, tattoos, and trinkets. In comics, Willie the Penguin had a short run from 1951 to 1952 and Chilly Willy picked up the action in 1953. Tennessee Tuxedo, accompanied by his pal Chumley the Walrus, had adventures in the 1960s. There is also Pokey the Penguin, an online comic strip. One of the most persistent characterizations is the Penguin of Batman comic books. The various incarnations of the Penguin in the Batman movies is about as nasty a character as ever was depicted in the comics or movies. The evil motives assigned to this penguin contrast sharply with the "cuteness" of most other images. The picture of Willie the Penguin on a cigarette package in about 1932 was intended to showcase that inhalation of menthol-treated tobacco was cool, and he remained their mascot for several decades. Times have changed; the recent Kool cigarette packaging no longer features a penguin.

Another figure that is never called a penguin, although it has some suspiciously similar features, is the Shmoo in the *Li'l Abner* comic strip of Al Capp. The Shmoo, shaped like a bowling pin with legs, has no arms, ears, or nose but does have whiskers. It reminds me of an emperor penguin chick, the only penguin that walks with its wings hard against its side; it appears wingless with its uniform gray down. All are attributes like those of the Shmoo. During the short run the Shmoo had, from 1949 to 1950, it was a great success both in the popularity of the strip and in commerce. There was even a book, *The Life and Times of the Shmoo.* To many others in science, the Shmoo has evoked various other items of similar shape. They range from the field of microbiology, in which it resembles budding yeast, through marine biology, in which echinoderm specialists refer to shmoos as nondescript larvae, to particle physics, in which it is a high-energy cosmic ray detector.

There are many books about penguins, and more are being written all the time by numerous authors. A good list of titles can be found on the New England Aquarium website, www.neaq.org/education_and_activities/student_resources/recommended_books/books_about_penguins.php. There is also a useful Wikipedia page, "List of Fictional Penguins" (http://en.wikipedia.org/wiki/List_of_fictional_penguins), that lists numerous fictional penguins in addition to those I have mentioned. At the top of the New England Aquarium's list are children's books on penguins. The most factual are the ones written by artists and writers who have visited the Antarctic under the special National Science Foundation program conceived and executed by Guy Guthridge (retired), who was an administrator in the Office of Polar Programs of the National Science Foundation.

One of those books is *My Season with Penguins: An Antarctic Journal*, by Sophie Webb, who was invited to Antarctica in 1996 to observe Adélie penguins. A biologist and an artist, she illustrates and tells a personal story of her two-month stay at an Adélie penguin colony near McMurdo Station, Antarctica.

A Mother's Journey, by Sandra Markle, was written after Markle was chosen to journey to Antarctica in 1996 and 1998 as part of the National Science Foundation Artists and Writers Program. To her credit, she took a different tack on emperor penguin breeding behavior. Instead of recounting the much-emphasized trial of the male, which fasts for four months while it incubates the egg, she tells the story of the female, making the trek back to sea after laying the egg. Before returning to the male and the chick, the mother penguin has the task of hunting for food during the winter conditions of intense cold and darkness.

A much earlier children's book of pure fiction was Richard and Florence Atwater's *Mr. Popper's Penguins*, published in 1938. It is the leader in a genre of books that pay little attention to the facts of the natural history of penguins. It tells about a pair of penguins and their "batch" of 12 chicks, which are kept by the Popper family. Naturally, a dozen penguins become an overpopulation problem in the Popper household, and eventually they are taken to the North Pole for release. Perhaps this, in part, has contributed to the belief by many that penguins live at the North Pole. In later years this inaccuracy was promoted by Gary Larson's famous cartoon of a polar bear that was disguised as a penguin while searching for a meal on an ice floe.

The Popper book was adapted for a full feature movie in 2011, but it is similar to the book only in its title. Now Mr. Popper is no longer a small-

Adélie penguin.

town painter, but a New York realtor living on Park Avenue. He converts his apartment into an ice palace for his collection of what are clearly gentoo penguins. The gentoo is the third largest of all penguins living now. The story concludes with the Popper family returning the birds to Antarctica.

In addition to movie dramas, there are several full-length cartoons. *Happy Feet*, complete with a sequel, has been the most successful. There is also *The Penguins of Madagascar*, an animated TV series that ran from 2009 to 2010. In this series a pack of African animals, which includes penguins, move from jungles and plains to New York City. Such a range of adventures offers unlimited exploits for the inventive cartoonist to create. Another animated cartoon is *The Pebble and the Penguin*, a 2008 love story and adventure. There is also the 2007 movie *Farce of the Penguins*, which, as far as I know, is the only animated feature film about penguins that is rated R. The scene is set in the Antarctic, and that is the end of its authenticity. There are others that could be listed, but the concept is clear: one of the main protagonists of modern-day fairy tales is the penguin.

Of a more serious nature in the documentary style is the 2005 *March of the Penguins*. The story line and emphasis are summarized in its introduction: "Every March since the beginning of time, the quest begins to find the perfect mate and start a family. This courtship will begin with a long journey—a journey that will take them hundreds of miles across the continent by foot, in freezing cold temperatures, in brittle, icy winds and through deep, treacherous waters. They will risk starvation and attack by dangerous predators, under the harshest conditions on earth, all to find true love." Narrated by Morgan Freeman, this film is the most successful wildlife documentary ever; it won an Oscar and made more money than any other animal movie. Also worth noting is *Crittercam. Emperor Penguins*, a penguin diving and feeding documentary that came out in a TV series called the *Crittercam Chronicles*, which was produced and directed by Greg Marshall for National Geographic. It is included as a feature in the DVD of *March of the Penguins*. *Of Men and Penguins*, the story of the photographers in the field making the movie, is also included on that DVD.

One of the earliest full-length documentaries shown on television, *Penguin City*, was filmed and produced in 1970 by W. J. L. Sladen. It is about the breeding biology of one of the largest Adélie penguin colonies in Antarctica. About forty years later, a sequel, *Return to Penguin City*, produced by David Ainley, a former student of Sladen, features the work of Grant Ballard and Viola Toniolo, two biologists from California. It explains that, "using advanced technology and old-school field research, they are discovering how Antarctic penguins—creatures that already survive on the edge—are coping with rapid climate change." The video was recorded at Cape Crozier, the same colony featured in the original documentary. Finally, a series of hour-long documentaries titled *The Frozen Planet*, produced by the British Broadcasting Company, features penguins in some episodes, and the imaging is the most beautiful and sensational captured so far. It sets a new technical standard of videography.

What roles have penguins played in poetry and other literature?

There is very little mention of penguins in poetry, except perhaps in recent literature, after the explosion of books, comics, and movies on penguins that began in 2000. Much of it has been prompted by the success of *March of the Penguins*.

Coffee-table books and other more technical books, written by professional photographers and presenting stunning images of penguins, are listed on the New England Aquarium website, www.neaq.org/education_and_activities/student_resources/recommended_books/books_about_penguins.php. In my view, the top three on the list for sound information on the natural history of penguins are *The Penguins* (part of a series, Bird Families of the World), by Tony Williams; *Penguins Past, Present, Here and There*, by George Gaylord Simpson; and *Penguins of the World*, by Wayne Lynch. Lynch is also the photographer for this book. In addition to being very informative about natural history in general, Lynch's book contains some of the best published photographs of penguins.

Chapter 12

"Penguinology"

Who studies penguins?

Because of the special attraction that humans feel for penguins, there are many people from different vocations that study penguins. Most are biologists, but in addition there are physicians, lawyers, physicists, and others conducting work on penguins. Some have converted completely to specializing in research on penguin biology. There are also countless authors and photographers who have completed at least one report or photo essay on penguins or have written books about them. I have joined the latter group, as well as being a longtime member of the group specializing in penguin research. This is my first book about penguins, after decades of study and publishing research papers in scientific journals.

Which species are best known?

Unquestionably the best-known penguin, in regard to the public in general, is the emperor penguin, which has two famous feature films to its credit, *March of the Penguins* and *Happy Feet.* The most-read chronicle of an expeditionary journey to observe a wild animal, the emperor penguin in winter, is Apsley Cherry-Garrard's book *Worst Journey in the World.* The best-known individual of all emperor penguins is one that strayed onto the north island of New Zealand in June 2011. Before he was released, after a four-month rehabilitation program, the news about him had spread worldwide, and he had been given the flippant name of "Happy Feet." Personally, I prefer the moniker "Endurance," because he was like Sir Ernest Henry Shackleton and his crew, who in the early twentieth century crossed

Emperor penguin adult with a group of chicks.

from Elephant Island to South Georgia Island during the winter. That journey was against all odds, and in a way so was Happy Feet's. Sadly, the final chapter in Happy Feet's story was not as heroic as Shackleton's. When Happy Feet was released just north of the Southern Ocean in September of the same year, his luck ran out and he disappeared four days after release.

Emperor penguins, I would wager, have appeared on the covers of more books and magazines than any other species of wildlife, including the giant panda. The two have a commonality despite their great differences, in that they are black and white. Well, mostly black and white: emperor penguins have a beautiful touch of yellow on the neck and chest, which pandas lack. A close second in penguin popularity is the other completely Antarctic species, the Adélie penguin. It too is exceptionally iconic. A smaller bird than the emperor penguin, the Adélie is more numerous and often the first Antarctic bird seen by travelers to the Antarctic. Both species have been the subjects of numerous documentary films and books, as well as a large number of research papers dealing with their behavior, physiology, and ecology. There is more known about the diving physiology of emperor penguins than probably any other air-breathing vertebrate except humans, thanks to the research efforts of Paul Ponganis, an anesthesiologist converted to penguin biology, and his students. To mention a few facts, the emperor penguin's heart rate while diving to depth has shown a wide range, from slight decreases from the resting heart rate of about 60 beats/minute to a rate as low as 6 beats/minute near the end of an 18 minute dive. Emperor penguins have an impressive tolerance for hypoxia (oxygen deficiency), so that their venous blood oxygen tension can decrease to as low as 6 mmHg at the end of a 23 minute dive. Such responses translate into a maximum breath-hold capacity of at least 27 minutes—truly astonishing for such a

Two Adélie penguins calling while in the process of exchanging the egg under the bird in the foreground with the bird in the background.

small diver, compared to the capacity of many marine mammals of much larger size.

Which species is least known?

Most likely the Snares penguin is the least known, or the least often seen. It lives on remote uninhabited islands of New Zealand. Tourists are not allowed on the Snares Island group, so the bird is unlikely to be seen unless it turns up on other islands, such as Stewart or South Island of New Zealand. It breeds on only four islands, all of them within the Snares Island group. There are only about 30,000 pairs, and they are difficult to recognize at sea. Because of their limited range and because they breed on only one small island group, Snares penguins are considered vulnerable by the International Union for the Conservation of Nature. However, trends in their population are unknown, and if they are discovered to be declining, the listing will probably be changed to mark them as endangered.

The Snares penguin breeds on a rocky island in one of the windiest places and surrounded by some of the most turbulent waters on the planet.

How do scientists tell penguins apart?

They are distinguished by location, for starters, because the various penguin species are in most cases widely separated geographically. This is especially true for the banded penguins. If you are in Galápagos and see a penguin, it will be a Galápagos penguin. If you see a penguin in South Africa, it has to be an African penguin. The subtleties of banding in the *Spheniscus* group of penguins is not important until you visit a zoo where there may be more than one species of banded penguin. Then it can be confusing. There is more likely an overlap among the *Eudyptes* or crested group of penguins at sea and on some of the southern islands around New Zealand. Some in this group tend to wander widely, based on records of sightings in New Zealand, Australia, and South Africa of the rockhopper penguin, and the Snares Island penguin has paid visits to the near neighborhoods of Australia and New Zealand. At sea it is almost impossible to tell the crested penguins apart, and the same is true of most other species within their sister group. Much identification is based on the probability arising from the location and abundance of the species. For example, if you are sailing near the Snares Island group, the chances are that the bird sighted is a Snares penguin, but it could also be a rockhopper or almost any one of the crested penguins that might visit the island. The most difficult penguin group of all for differentiation at sea would be the rockhopper and the macaroni around the tip of South America, the Falklands, Crozet Archipelago, and Kerguelen, where both species are abundant. Similarly, errors can be made easily between chinstrap and Adélie penguins in the Antarctic Peninsula area. Of course, on land chinstrap and Adélie penguins are easy to tell apart. Most difficult of all to distinguish on land or at sea are the three subspecies of rockhoppers.

If there is a need to tell individuals apart, then the only means for long-term studies are either transponders injected under the skin or external flipper bands. They both have their advantages and disadvantages. Transponders have a range of only a few centimeters, so that the receiver must be close to the penguin. Such a limited range is no problem for species that have specific nest sites that must be reached along a narrow path. The transponder platform reader is placed underground or below the snow path leading to the only entrance into the colony. An identification code is transmitted to the receiver in the reader. Under special conditions, this system has worked well for king and Adélie penguins. The implant will last the life of the penguin, with little or no evidence of serious side effects. Bands are of exceptional value because they can be seen from a distance and the individual identified. Unfortunately, they induce mortality in a percentage of the birds banded and there is a reluctance to use them at present, but they have been used widely in earlier years before all of the problems were well known. They are still used in some species, but the justification for their use is more critically considered. It is also necessary to evaluate whether the negative effects associated with one banded species are likely to apply to another.

Appendix A

Penguins of the World

Class Aves
Order Sphenisciformes
Family Spheniscidae

Scientific name	Common name	General location	Food
Aptenodytes forsteri	emperor penguin	All coastal Antarctic	fish, squid, krill
Aptenodytes patagonicus	king penguin	All sub-Antarctic islands	lanternfish
Pygoscelis adélie	Adélie penguin	All coastal Antarctic and many Antarctic islands	krill, fish
Pygoscelis antarctica	chinstrap penguin	Antarctic Peninsula and South Shetland, Sandwich, Orkney, and Balleny Islands	krill
Pygoscelis papua	gentoo penguin	Antarctic Peninsula, all sub-Antarctic islands, and South Shetland Islands	fish, krill
Eudyptula minor	little penguin	New Zealand and South Australia	inshore, pelagic fish
Spheniscus demersus	African penguin	South Africa and Namibia	anchovy
Spheniscus magellanicus	Magellanic penguin	Chile and Argentina	anchovy and sardine
Spheniscus humboldti	Humboldt penguin	Chile and Peru	anchovy and sardine
Spheniscus mendiculus	Galápagos penguin	Galápagos Islands, Ferdinandina and Isabela Islands	sardine and anchovy
Megadyptes antipodes	yellow-eyed penguin	New Zealand, SE South Island, Stewart, Auckland, and Campbell Islands	varied pelagic and demersal fish and squid
Eudyptes pachyrhynchus	fiordland penguin	Southwest New Zealand	squid and ?
Eudyptes robustus	Snares penguin	Snares Island	unknown; mostly euphausids
Eudyptes sclateri	erect crested penguin	Bounty and Antipodes Islands	Unknown
Eudyptes chrysocome	rockhopper penguin	All sub-Antarctic islands and Tristan da Cunha, Bounty, and Antipodes Islands	euphausids, fish, squid
Eudyptes chrysolophus	macaroni penguin	Southern islands of Argentina and Chile, Falkland Islands, islands of the Scotia Arc from South Georgia to South Shetlands, Bouvet, Prince Edward Islands, Crozet Islands, Kerguelen Islands, Heard and McDonald Islands	krill
Eudyptes schlegeli	royal penguin	Macquarie Island	euphausids and lanternfish

Appendix B

Penguin Research and Conservation Organizations

Below are some of the most prominent conservation and research organizations. Several of them operate in the Southern Hemisphere where penguins live, and others are government research organizations that make a substantial contribution to penguin research.

Southern African Foundation for the Conservation of Coastal Birds

The Southern African Foundation for the Conservation of Coastal Birds was established near Cape Town in 1968 in response to oil spills in the Cape Town area. Its specialty is rehabilitation of oiled African penguins, and it has treated more than 90,000 seabirds. After the *MV Treasure* oil spill, staff of this organization handled more than 38,000 penguins, half of which had been oiled. See SANCCOB, www.sanccob.co.za/?m=1.

Penguin Conservation Centre

The Penguin Conservation Centre, near Cape Town, South Africa, was established in 1968 to care for oiled seabirds. It is a multispecies conservation organization with projects for volunteers on animals ranging from penguins to leopards, but the organization's specialty is African penguins and their rehabilitation. It sponsors a variety of projects in which volunteers may stay and work for various lengths of time. See Enkosini Eco Experience, www.enkosiniecoexperience.com/PenguinConservationCentre.htm.

Magellanic Penguin Project

The Magellanic Penguin Project was established jointly in 1982 between the Wildlife Conservation Society, the Province of Chubut, and the University of Washington. The catalyst for its conception was a concession sought by a company to harvest penguins. At the time, about 20,000 adults and 22,000 juveniles were dying each year from oil pollution caused by tankers dumping ballast water. By 1997 the Argentine government moved the tanker lanes 40 km farther offshore. This greatly reduced the oiling problem in the local area, but large numbers of Magellanic penguins continue to die of oil pollution. The colony is also a hotspot for tourism; about 62,000 people visited the Punta Tombo colony in 2005–6. Research and conservation projects have been continuous for about 40 years. See Penguin Sentinels, http://mesh.biology.washington.edu/penguinProject/home.

Humboldt Penguin Conservation Centre

The Humboldt Penguin Conservation Centre is a project of the World Association of Zoos and Aquariums, whose goal is to facilitate changing the designation of Punta San Juan (PSJ) from a guano mine to a marine reserve. This move is badly needed. The industrial fishery off the coast of Peru is the largest in the world. The average annual take is 8 to 12 million metric tons, or about 10% of the global marine catch, and the competition of the industry with other marine predators has resulted since the 1950s in a decline from 28 million to 1.8 million guanay. The mined guano was once an important nesting habitat for the Humboldt penguin. The Humboldt penguin population at PSJ is only about 2,500 pairs, small compared to the numbers of some other species. This penguin is in decline and needs to have a formal organization supporting its conservation. See World Association of Zoos and Aquariums, www.waza.org/en/site/conservation/waza-conservation-projects/overview/humboldt-penguin-conservation-centre.

Yellow-Eyed Penguin Trust

The Yellow-Eyed Penguin Trust was formed in 1987 by conservationists in Dunedin, New Zealand. Their major goal was to restore habitat for the yellow-eyed penguin and control predators. The present tasks of the trust are revegetation, predator trapping, and education. See Yellow-Eyed Penguin Trust, http://yellow-eyedpenguin.org.nz/.

Antarctic Krill Conservation Project

The Antarctic Krill Conservation Project is guided by a steering committee consisting of the Antarctic and Southern Ocean Coalition and the Pew Charitable Trusts. It is one of several conservation groups that have been formed to conserve the Antarctic ecosystem and food web through the trophic levels from krill to whales. Another associated group is Friends of the Ross Sea Ecosystem. This group recognizes that the Antarctic, and especially the Ross Sea, represent the last chance for a pristine ocean environment to remain intact. Some of the most pressing issues are the use of krill for aquaculture and also its use by humans as a source of omega-3 fatty acids. There is concern about a special technology for rapidly harvesting large amounts of krill, which is the food base for almost all Antarctic predators. The harvesting of toothfish (aka Antarctic cod and Chilean sea bass) in the Ross Sea causes concern because little is known about the natural history of the animal, especially its reproductive biology. There is evidence that the population found in McMurdo Sound has gone into sharp decline since the harvest began. Also, with the climate-warming trend now in progress in the Southern Ocean, there is a decline in sea ice extent. Krill reproduction is dependent on sea ice, because the larval krill rise from the deep after hatching, and from the most advanced larval stage to adulthood, they graze on algae beneath and attached to the pack ice. See "Antarctic Krill Conservation Project," www.krillcount.org/index.html.

U.S. Antarctic Program

The U.S. Antarctic Program, sponsored by the National Science Foundation Office of Polar Penguins, maintains three year-round research stations in Antarctica, two of which are the bases for research scientists studying penguins. McMurdo Station, the oldest, was established in 1955. It is the largest base in Antarctica and is located at 77°51'S, 166°40'E. It and its neighbor Scott Base (sponsored by the New Zealand government) are the farthest-south coastal bases in Antarctica. McMurdo Station is near both emperor and Adélie penguin colonies and supports research on all western Ross Sea emperor penguin colonies and on the three nearest Adélie penguin colonies. In the Antarctic Peninsula is the Palmer Station laboratory, near several chinstrap, gentoo, and Adélie penguin colonies. Two Adélie and chinstrap penguin colonies have benefited from more than 30 years of long-term studies by scientists working from this laboratory.

Centre National de la Recherche Scientifique

The Centre National de la Recherche Scientifique sponsors the program of Terres australes et antarctiques françaises, which has one base in the Antarctic and two bases in the sub-Antarctic. The Antarctic base at Dumont d'Urville is where all emperor penguin studies have been conducted since 1958, and the scientists there have conducted long-term studies of the population since the inception of the base. The result is the longest population study of emperor penguins. Some of the study results have created sensations in the popular press because of substantial declines in the colony in 1976 and 1978. This is also where *March of the Penguins* was filmed. The sub-Antarctic main base, Alfred Faure, is on Île de la Possession, in the

Crozet archipelago. It has been continuously manned since 1963. One of this facility's main projects is long term studies of king penguins. This is the only place where killer whales are often seen taking penguins; the prey are king penguins. It was the only place where such activity was seen at all, until recent observations of killer whales taking gentoo and chinstrap penguins in the Antarctic Peninsula. There are also rockhopper penguin and macaroni penguins on the island. See Centre national de la recherche scientifique, http://www.cnrs.fr/.

British Antarctic Survey

For the British Antarctic Survey, the seed of the present program started in 1943 and was called Operation Tabarin. Its goal was to establish permanent bases in the Antarctic. At the end of World War II, the bases were renamed the Falkland Island Dependencies Survey, and the Survey received its current name in 1962. The program has three Antarctic research stations and two sub-Antarctic research stations. Most penguin studies are conducted at Signy Research Station on Signy Island, where the three brush-tailed penguins live. There is also some research accomplished at Halley Research Station, which has an emperor penguin colony nearby. There are numerous avian and fur seal studies conducted at the Bird Island Research Station on Bird Island, and occasional king penguin studies are conducted on nearby South Georgia Island. See British Antarctic Survey, http://www.antarctica.ac.uk/.

Bibliography

Adams, N. J. 1987. Foraging range of king penguins *Aptenodytes patagonicus* during summer at Marion Island. *Journal of Zoology* 212:475–482.

Ainley, D. G. 2002. *The Adélie Penguin, Bellwether of Climate Change.* 1st ed. New York: Columbia University Press.

Ainley, D. G., and G. Ballard. 2012. Non-consumptive factors affecting foraging patterns in Antarctic penguins: A review and synthesis. *Polar Biology* 35:1–13.

Ainley, D., J. Russell, S. Jenouvrier, E. Woehler, P. O. Lyver, W. R. Fraser, and G. L. Kooyman. 2010. Antarctic penguin response to habitat change as Earth's troposphere reaches 2 degrees C above preindustrial levels. *Ecological Monographs* 80:49–66.

Ancel, A., M. Horning, and G. L. Kooyman. 1997. Prey ingestion revealed by oesophagus and stomach temperature recordings in cormorants. *Journal of Experimental Biology* 200:149–154.

Ancel, A., H. Visser, Y. Handrich, D. Masman, and Y. LeMaho. 1997. Energy saving in huddling penguins. *Nature* 385:304–305.

Aubin, T., P. Jouventin, and C. Hildebrand. 2000. Penguins use the two-voice system to recognize each other. *Proceedings of the Royal Society of London, Series B, Biological Sciences* 267:1081–1087.

Baker, A. J., S. L. Pereira, O. P. Haddrath, and K. A. Edge. 2006. Multiple gene evidence for expansion of extant penguins out of Antarctica due to global cooling. *Proceedings of the Royal Society, Series B, Biological Sciences* 273:11–17.

Bannasch, R. 1995. Hydrodynamics of penguins: An experimental approach. In *The Penguins: Ecology and Management: Second International Penguin Conference, Cowes, Victoria, Australia, August 1992*, ed. P. Dann, I. Norman, and P. Reilly, 141–176. Chipping Norton, New South Wales, Australia: Surrey Beatty and Sons.

Barber, R. T., and F. P. Chavez. 1986. Ocean variability in relation to living resources during the 1982–83 El-Niño. *Nature* 319:279–285.

Baroni, C., and G. Orombelli. 1994. Abandoned penguin rookeries as Holocene paleoclimatic indicators in Antarctica. *Geology* 22:23–26.

Blainey, Geoffrey. 1976. *Triumph of the Nomads: A History of Aboriginal Australia.* New York: Overlook Press.

Boersma, P. D. 1978. Breeding patterns of Galapagos penguins as an indicator of oceanographic conditions. *Science* 200:1481–1483.

Bost, C. A., J. B. Charrassin, Y. Clerquin,Y. Ropert-Coudert, and Y. Le Maho. 2004. Exploitation of distant marginal ice zones by king penguins during winter. *Marine Ecology Progress Series* 283:293–297.

Burger, A., and M. Simpson. 1986. Diving depths of atlantic puffins and common murres. *Auk* 103:828–830.

Burger, A., R. P. Wilson, D. Garnier, and M-P. T. Wilson. 1993. Diving depths, diet, and underwater foraging of Rhinoceros Auklets in British Columbia. *Canadian Journal of Zoology* 71:2528–2540.

Cannell, B. L., and J. M. Cullen. 1998. The foraging behaviour of Little Penguins *Eudyptula minor* at different light levels. *Ibis* 140:467–471.

Challet, E., C. A. Bost, Y. Handrich, J. P. Gendner, and Y. Le Maho. 1994. Behavioural time budget of breeding king penguins (*Aptenodytes patagonica*). *Journal of Zoology (London)* 233:669–681.

Chappell, M. A., and S. L. Souza. 1988. Thermoregulation, gas exchange, and ventilation in Adelie penguins (*Pygoscelis adeliae*). *Journal of Comparative Physiology B* 157:783–790.

Charrassin, J. B., C. A. Bost, K. Puetz, J. Lage, T. Dahier, T. Zorn, and Y. Le Maho. 1998. Foraging strategies of incubating and brooding king penguins *Aptenodytes patagonicus*. *Oecologia (Berlin)* 114:194–201.

Cherel, Y., and G. L. Kooyman. 1998. Food of emperor penguins (*Aptenodytes forsteri*) in the western Ross Sea, Antarctica. *Marine Biology (Berlin)* 130:335–344.

Cherel, Y., K. Pütz, and K. A. Hobson. 2002. Summer diet of king penguins (*Aptenodytes patagonicus*) at the Falkland Islands, southern Atlantic Ocean. *Polar Biology* 25:898–906.

Cherel, Y., and V. Ridoux. 1992. Prey species and nutritive value of food fed during summer to King penguin *Aptenodytes patagonicus* chicks at Possession Island, Crozet Archipelago. *Ibis* 134:118–127.

Cherel, Y., C. Verdon, and V. Ridoux. 1993. Seasonal importance of oceanic myctophids in king penguin diet at Crozet Islands. *Polar Biology* 13:355–357.

Cherel, Y., and H. Weimerskirch. 1995. Seabirds as indicators of marine resources: Black-browed albatrosses feeding on ommastrephid squids in Kerguelen waters. *Marine Ecology Progress Series* 129:295–300.

Cherry-Garrard, A. 1922. *The Worst Journey in the World, Antarctic 1910–1913*. 1st ed. London: Constable and Company.

Clark, B. D., and W. Bemis. 1979. Kinematics of swimming of penguins at the Detroit Zoo. *Journal of Zoology (London)* 188:411–428.

Clarke, J., K. Knowles, C. Fowler, R. Lawless, S. Eberhard, and R. Murphy. 2003. Post-fledging and winter migration of Adélie penguins *Pygoscelis adéliae* in the Mawson region of East Antarctica. *Marine Ecology Progress Series* 248:267–278.

Clarke, J. A., D. T. Ksepka, M. Stucchi, M. Urbina, N. Giannini, S. Bertelli, Y. Narvàez, and C. A. Boyd. 2007. Paleogene equatorial penguins challenge the proposed relationship between biogeography, diversity, and Cenozoic climate change. *Proceedings of the National Academy of Sciences* 104:11545–11550.

Clottes, J. 2001. France's magical ice age art. *National Geographic* 200 (2): 104–122.

Clottes, J., and J. Courtin. 1993. Neptune's Ice Age gallery. *Natural History* 102:64–72.

Cott, H. B. 1985. Coloration, adaptive. In *A Dictionary of Birds*, ed. B. Campbell and E. Lack, 97–100. Shipman, VA: Buteo Books Imprint.

Crawford, R. J. M. 1998. Responses of African penguins to regime changes of sardine and anchovy in the Benguela system. *South African Journal of Marine Science* 19:355–364.

Crawford, R. J. M., R. Altwegg, B. J. Barham, P. J. Barham, J. M. Durant, B. M. Dyer, D. Geldenhuys, A. B. Makhado, L. Pichegru, P. G. Ryan, L. G. Underhill, L. Upfold, J. Visagie, L. J. Waller, and P. A. Whittington. 2011. Collapse of South Africa's penguins in the early 21st century. *African Journal of Marine Science* 33:139–156.

Croll, D., A. J. Gaston, A. E. Burger, and D. Konnoff. 1992. Foraging behavior and physiological adaptation for diving in thick-billed murres. *Ecology* 73:344–356.

Croxall, J. P., and G. S. Lishman. 1987. The food and feeding ecology of penguins.

In *Seabirds Feeding Ecology and Role in Marine Ecosystems*, ed. J. P. Croxall, 101–131. Cambridge: Cambridge University Press.

Culik, B., J. Hennicke, and T. Martin. 2000. Humboldt penguins out manouvering El Niño. *Journal of Experimental Biology* 203:2311–2322.

Culik, B. M., and G. Luna-Jorquera. 1997. The Humboldt penguin *Spheniscus humboldti*: A migratory bird? *Journal für Ornithologie* 138:325–330.

Darby, J. T., and S. M. Dawson. 2000. Bycatch of yellow-eyed penguins (*Megadyptes antipodes*) in gillnets in New Zealand waters, 1979–1997. *Biological Conservation* 93:327–332.

Davenport, J., R. N. Hughes, M. Shorten, and P. S. Larsen. 2011. Drag reduction by air release promotes fast ascent in jumping emperor penguins—a novel hypothesis. *Marine Ecology, Progress Series* 430:171–182.

Davis, L., R. Harcourt, and C. Bradshaw. 2001. The winter migration of Adelie penguins breeding in the Ross Sea sector of Antarctica. *Polar Biology* 24:593–597.

Davis, L. S., and M. Renner. 2003. *Penguins.* 1st ed. New Haven, CT: Yale University Press.

Deagle, B. E., N. J. Gales, and M. A. Hindell. 2008. Variability in foraging behaviour of chick-rearing macaroni penguins *Eudyptes chrysolophus* and its relation to diet. *Marine Ecology, Progress Series* 359:295–309.

Diamond, J., and A. B. Bond. 2003. A comparative analysis of social play in birds. *Behaviour* 140:1091–1115.

Dobson, F. S., C. Couchoux, and P. Jouventin. 2011. Sexual selection on a coloured ornament in king penguins. *Ethology* 117:872–879.

Dobson, F. S., P. M. Nolan, M. Nicolaus, C. Bajzak, A. S. Coquel, and P. Jouventin. 2008. Comparison of color and body condition between early and late breeding king penguins. *Ethology* 114:925–933.

Drent, R. H., and B. Stonehouse. 1971. Thermoregulatory responses of the Peruvian penguin, *Spheniscus humboldti*. *Comparative Biochemistry and Physiology* 40A:689–710.

Dresp, B., P. Jouventin, and K. Langley. 2005. Ultraviolet reflecting photonic microstructures in the King Penguin beak. *Biology Letters* 1:310–313.

Duignan, P. 2001. Diseases of penguins. *Surveillance* 28:5–11.

Dunn, M. J., J. R. D. Silk, and P. N. Trathan. 2011. Post-breeding dispersal of Adélie penguins (*Pygoscelis adéliae*) nesting at Signy Island, South Orkney Islands. *Polar Biology* 34:205–214.

Ellegren, H., and B. C. Sheldon. 1997. New tools for sex identification and the study of sex allocation in birds. *Trends in Ecology and Evolution* 12:255–259.

Fretwell, P. T., M. A. LaRue, P. Morin, G. L. Kooyman, B. Wienecke, N. Ratcliffe, A. J. Fox, A. H. Fleming, C. Porter, and P. N. Trathan. 2012. An emperor penguin population estimate: The first global, synoptic survey of a species from space. *PLoS ONE* 7:e33751.

Gamble, J. R., and D. A. Cristol. 2002. Drop-catch behavior is play in herring gulls, *Larus argentatus*. *Animal Behaviour* 63:339–345.

Garcia-Borboroglu, P., P. D. Boersma, V. Ruoppolo, R. Pinho-da-Silva, A. Corrado-Adornes, D. Conte-Sena, R. Velozo, C. Myiaji-Kolesnikovas, G. Dutra, P. Maracini, C. Carvalho-do-Nascimento, V. Ramos, L. Barbosa, and S. Serra. 2010. Magellanic penguin mortality in 2008 along the SW Atlantic coast. *Marine Pollution Bulletin* 60:1652–1657.

Gilbert, C., Y. Le Maho, M. Perret, and A. Ancel. 2007. Body temperature changes induced by huddling in breeding male emperor penguins. *American Journal of Physiology: Regulatory Integrative and Comparative Physiology* 292:R176–R185.

Gilbert, C., G. Robertson, Y. Le Maho, Y. Naito, and A. Ancel. 2006. Huddling behavior in emperor penguins: Dynamics of huddling. *Physiology and Behavior* 88:479–488.

Graham, J. B. 1974. Aquatic respiration in the sea snake. *Respiration Physiology* 21:1–7.

Green, K., and R. G. M. Williams. 1998. Foraging ecology and diving behaviour of Macaroni Penguins *Eudyptes chrysolophus* at Heard Island. *Marine Ornithology* 26:27–34.

Griffin, T. M., and R. Kram. 2000. Biomechanics: Penguin waddling is not wasteful. *Nature* 408:929.

Hagelin, J. C. 1992. The structure and function of penguin feathers. Unpublished. In Hagelin's personal collection.

Handrich, Y., J. P. Gendner, and Y. Le Maho. 1996. Breeding penguins as indicators of marine resources: A study with minimal human disturbance. In *The Penguins: Ecology and Management: Second International Penguin Conference, Cowes, Victoria, Australia, August 1992*, ed. P. I. N. Dann and P. Reilly, 72–79. Chipping Norton, New South Wales, Australia: Surrey Beatty and Sons.

Heinrich, B. 1999. *Mind of the Raven: Investigation and Adventures with Wolf-Birds*. New York: Harper Collins.

Herling, C., B. M. Culik, and J. C. Hennicke. 2005. Diet of the Humboldt penguin (*Spheniscus humboldti*) in northern and southern Chile. *Marine Biology* 147:13–25.

Hinke, J. T., K. Salwicka, S. G. Trivelpiece, G. M. Watters, and W. Z. Trivelpiece. 2007. Divergent responses of Pygoscelis penguins reveal a common environmental driver. *Oecologia* 153:845–855.

Howland, H. C., and J. G. Sivak. 1984. Penguin vision in air and water. *Vision Research* 24:1905–1909.

Hsieh, C.-h., C. S. Reiss, J. R. Hunter, J. R. Beddington, R. M. May, and G. Sugihara. 2006. Fishing elevates variability in the abundance of exploited species. *Nature* 443:859–62.

Hull, C. L. 2000. Comparative diving behaviour and segregation of the marine habitat by breeding Royal Penguins, *Eudyptes schlegeli*, and eastern Rockhopper Penguins, *Eudyptes chrysocome filholi*, at Macquarie Island. *Canadian Journal of Zoology* 78:333–345.

Hunter, F. M., and L. S. Davis. 1998. Female Adélie Penguins acquire nest material from extrapair males after engaging in extrapair copulations. *Auk* 115:526–528.

Idyll, C. P. 1973. The anchovy crises. *Scientific American* 228:22–29.

Isaacs, J., ed. 2006. *Australian Dreaming: 40,000 Years of Aboriginal History*. Sydney, Australia: New Holland Publishers.

Jansen, J. K., P. L. Boveng, and J. L. Bengtson. 1998. Foraging modes of chinstrap penguins: Contrasts between day and night. *Marine Ecology Progress Series* 165:161–172.

Jouventin, P. 1975. Mortality parameters in emperor penguin *Aptenodytes forsteri*. In *The Biology of Penguins*, ed. B. Stonehouse, 435–446. London: Macmillan.

Jouventin, P., T. Aubin, and T. Lengagne. 1999. Finding a parent in a king penguin: The acoustic system of individual recognition. *Animal Behaviour* 57:1175–1183.

Jouventin, P., P. M. Nolan, F. S. Dobson, and M. Nicolaus. 2008. Coloured patches influence pairing rate in King Penguins. *Ibis* 150:193–196.

Jouventin, P., P. M. Nolan, J. Ornborg, and F. S. Dobson. 2005. Ultraviolet beak spots in King and Emperor penguins. *Condor* 107:144–150.

Kirkwood, R. 2001. Emperor Penguin (*Aptenodytes forsteri*) foraging ecology. Department of Environment and Heritage, Australian Antarctic Division, Kingston.

Kirkwood, R., and G. Robertson. 1997. The foraging ecology of female Emperor Penguins in winter. *Ecological Monographs* 67:155–176.

Kirkwood, R., and G. Robertson. 1997. Seasonal change in the foraging ecology of emperor penguins on the Mawson Coast, Antarctica: Travel, location, and habitat selection. *Marine Ecology Progress Series* 156:205–223.

Kooyman, G. 2002. Evolutionary and ecological aspects of some Antarctic and sub-Antarctic penguin distributions. *Oecologia* 130:485–495.

Kooyman, G. L., Y. Cherel, Y. Le Maho, J. P. Croxall, P. H. Thorson, V. Ridoux, and C. A. Kooyman. 1992. Diving behavior and energetics during foraging cycles in king penguins. *Ecological Monographs* 62 (1):143–163.

Kooyman, G. L., D. Croll, S. Stone, and S. Smith. 1990. Emperor penguin colony at Cape Washington, Antarctica. *Polar Record* 26:103–108.

Kooyman, G. L., and R. W. Davis. 1987. Diving behavior and performance, with special reference to penguins. In *Seabirds: Feeding Biology and Role in Marine Ecosystems*, ed. J. P. Croxall, 63–75. London: Cambridge University Press.

Kooyman, G., E. Hunke, S. Ackley, R. van Dam, and G. Robertson. 2000. Moult of the emperor penguin: Travel, location, and habitat selection. *Marine Ecology Progress Series* 204:269–277.

Kooyman, G. L., and T. G. Kooyman. 1995. Diving behavior of emperor penguins nurturing chicks at Coulman Island, Antarctica. *Condor* 97:536–549.

Kooyman, G. L., T. G. Kooyman, M. Horning, and C. A. Kooyman. 1996. Penguin dispersal after fledging. *Nature* 383:397.

Kooyman, G. L., and P. J. Ponganis. 1994. Emperor penguin oxygen consumption, heart rate, and plasma lactate levels during graded swimming exercise. *Journal of Experimental Biology* 195:199–209.

Kooyman, G. L., and P. J. Ponganis. 1997. The challenges of diving to depth. *American Scientist* 85:530–539.

Kooyman, G. L., and P. J. Ponganis. 2007. The initial journey of juvenile emperor penguins. *Aquatic Conservation: Marine and Freshwater Ecosystems* 17:S37–S43.

Kooyman, G., P. Ponganis, M. Castellini, E. Ponganis, K. Ponganis, P. Thorson, S. A. Eckert, and Y. LeMaho. 1992. Heart rates and swim speeds of emperor penguins diving under sea ice. *Journal of Experimental Biology* 165:161–80.

Kooyman, G. L., D. Siniff, I. Stirling, and J. Bengtson. 2004. Moult habitat, pre- and post-moult diet, and post-moult travel of Ross Sea emperor penguins. *Marine Ecology Progress Series* 267:281–290.

Kramer, M. O, 1965. Hydrodynamics of the dolphin. In *Advances in Hydroscience*, vol. 2, pp. 111–130, V.T. Chow, Ed. New York: Academic Press.

Ksepka, D. T., S. Bertelli, and N. P. Giannini. 2006. The phylogeny of the living and fossil Sphenisciformes (penguins). *Cladistics* 22:412–441.

Lefebvre, L., N. Nicolakakis, and D. Boire. 2002. Tools and brains in birds. *Behaviour* 139:939–973.

Le Maho, Y. 1977. Emperor penguin: Strategy to live and breed in cold. *American Scientist* 65:680–693.

Le Maho, Y., J. P. Gendner, E. Challet, C. A. Bost, J. Gilles, C. Verdon, C. Plumere, J. P. Robin, and Y. Handrich. 1993. Undisturbed breeding penguins as indicators of changes in marine resources. *Marine Ecology, Progress Series* 95:1–6.

Lourandos, H. 1997. *Continent of Hunter-Gatherers: New Perspectives in Australian Prehistory*. Cambridge: Cambridge University Press.

Low, P. S., S. S. Shabk, T. J. Sejnowski, and D. Margoliash. 2008. Mammalian-like features of sleep structure in zebra finches. *Proceedings of the National Academy of Sciences* 105:9081–9086.

Lyamin, O. I., P. R. Manger, S. H. Ridgway, L. M. Mukhametov, and J. M. Siegel. 2008. Cetacean sleep: An unusual form of mammalian sleep. *Neuroscience and Biobehavioral Reviews* 32:1451–1484.

Martin, G. R. 1999. Eye structure and foraging in King Penguins *Aptenodytes patagonicus*. *Ibis* 141:444–450.

Mattern, T., U. Ellenberg, D. M. Houston, and L. S. Davis. 2007. Consistent foraging routes and benthic foraging behaviour in yellow-eyed penguins. *Marine Ecology Progress Series* 343:295–306.

McClintock, J., H. Ducklow, and W. Fraser. 2008. Ecological responses to climate change on the Antarctic Peninsula. *American Scientist* 96:302–310.

Mills, K. 2000. Diving behaviour of two Galapagos penguins *Spheniscus mendiculus*. *Marine Ornithology* 28:75–79.

Moore, G. J., B. Wienecke, and G. Robertson. 1999. Seasonal change in foraging areas and dive depths of breeding king penguins at Heard Island. *Polar Biology* 21:376–384.

Mora, C., R. A. Myers, M. Coll, S. Libralato, T. J. Pitcher, R. U. Sumaila, D. Zeller, R. Watson, K. J. Gaston, and B. Worm. 2009. Management effectiveness of the world's marine fisheries. *PLoS Biology* 7 (6): e1000131.

Moreno, J., J. Bustamante, and J. Vinuela. 1995. Nest maintenance and stone theft in the Chinstrap penguin (*Pygoscelis antarctica*): 1. Sex roles and effects on fitness. *Polar Biology* 15:533–540.

Mougin, J.-L., and M. van Beveren. 1979. Structure et dynamique de la population de manchots empereur *Aptenodytes forsteri* de la colonie de l'archipel de Pointe Geologie, Terre Adelie. *Compte Rendus Academie Science de Paris* 289D:157–160.

Muller-Schwarze, D. 1978. Play behavior in the Adélie penguin. In *Evolution of Play Behavior*, vol. 47, ed. D. Mueller-Schwarze, 375–377. Stroudsburg, PA: Dowden, Hutchinson, and Ross.

Nicolaus, M., C. Le Bohec, P. M. Nolan, M. Gauthier-Clerc, Y. Le Maho, J. Komdeur, and P. Jouventin. 2007. Ornamental colors reveal age in the king penguin. *Polar Biology* 31:53–61.

Nolan, P. M., F. S. Dobson, B. Dresp, and P. Jouventin. 2006. Immunocompetence is signalled by ornamental colour in king penguins, *Aptenodytes patagonicus*. *Evolutionary Ecology Research* 8:1325–1332.

Nolan, P. M., F. S. Dobson, M. Nicolaus, T. J. Karels, K. J. McGraw, and P. Jouventin. 2010. Mutual mate choice for colorful traits in king penguins. *Ethology* 116:635–644.

Nolan, P. M., M. Nicolaus, C. Bajzak, A. S. Coquel, and P. Jouventin. 2006. Ornamental colors signal sex, health, and breeding status in king penguins, *Aptenodytes patagonicus*. *Integrative and Comparative Biology* 46:E104.

Patterson, E. M., and J. Mann. 2011. The ecological conditions that favor tool use and innovation in wild bottlenose dolphins (*Tursiops* sp.). *PLoS ONE* 6:e22243.

Peters, G. 1997. A new device for monitoring gastric pH in free-ranging animals. *American Journal of Physiology: Gastrointestinal and Liver Physiology* 273:G748–G753.

Peters, G., R. P. Wilson, J. A. Scolaro, S. Laurenti, J. Upton, and H. Galleli. 1998. The diving behavior of Magellanic Penguins at Punta Norte, Peninsula Valdes, Argentina. *Colonial Waterbirds* 21:1–10.

Pincemy, G., F. S. Dobson, and P. Jouventin. 2009. Experiments on colour ornaments and mate choice in king penguins. *Animal Behaviour* 78:1247–1253.

Pinshow, B., M. A. Fedak, D. R. Battles, and K. Schmidt-Nielsen. 1976. Energy expenditure for thermoregulation and locomotion in Emperor Penguins. *American Journal. of Physiology* 231:903–912.

Pitman, R., and J. Durban. 2011. Killer whale predation on penguins in Antarctica. *Polar Biology* 33:1589–1594.

Ponganis, P. J. 2007. Returning on empty: Extreme blood O2 depletion underlies dive capacity of emperor penguins. *Journal of Experimental Biology* 210:4279–4285.

Ponganis, P. J. 2011. Diving mammals. *American Physiological Society: Comparative Physiology* 1:447–465.

Ponganis, P. J., and G. L. Kooyman. 1991. Diving physiology of penguins. *Symposium 33: Physiology of diving birds*. In *ACTA XX Congressus Internationalis Ornithologici*. Christchurch, New Zealand: Ornithological Congress Trust Board.

Ponganis, P. J., G. L. Kooyman, and S. H. Ridgway. 2003. Comparative diving physiology. In *Bennett and Elliott's Physiology and Medicine of Diving*, ed. A. O. Brubakk and T. S. Neuman, 211–226. Edinburgh: Saunders.

Ponganis, P. J., J. U. Meir, and C. L. Williams. 2010. Oxygen store depletion and the aerobic dive limit in emperor penguins. *Aquatic Biology* 8:237–245.

Ponganis, P. J., T. K. Stockard, J. U. Meir, C. L. Williams, K. V. Ponganis, and R. Howard. 2009. O_2 store management in diving emperor penguins. *Journal of Experimental Biology* 212:217–224.

Proctor, N. S., and P. J. Lynch. 1993. *Manual of Ornithology: Avian Structure and Function*. New Haven, CT: Yale University Press.

Putz, K., and C. A. Bost. 1994. Feeding behavior of free-ranging king penguins (*Aptenodytes patagonicus*). *Ecology (Tempe)* 75:489–497.

Putz, K., and Y. Cherel. 2005. The diving behaviour of brooding king penguins (*Aptenodytes patagonicus*) from the Falkland Islands: Variation in dive profiles and synchronous underwater swimming provide new insights into their foraging strategies. *Marine Biology* 147:281–290.

Putz, K., R. J. Ingham, J. G. Smith, and B. H. Luthi. 2002. Winter dispersal of rockhopper penguins *Eudyptes chrysocome* from the Falkland Islands and its implications for conservation. *Marine Ecology Progress Series* 240:273–284.

Putz, K., A. R. Rey, A. Schiavini, A. P. Clausen, and B. H. Luthi. 2006. Winter migration of rockhopper penguins (*Eudyptes c. chrysocome*) breeding in the South west Atlantic: Is utilisation of different foraging areas reflected in opposing population trends? *Polar Biology* 29:735–744.

Putz, K., A. Schiavini, A. R. Rey, and B. H. Luthi. 2007. Winter migration of magellanic penguins (*Spheniscus magellanicus*) from the southernmost distributional range. *Marine Biology* 152:1227–1235.

Putz, K., R. P. Wilson, J. B. Charrassin, T. Raclot, J. Lage, Y. Le Maho, M. A. M. Kierspel, B. M. Culik, and D. Adelung. 1998. Foraging strategy of King Penguins (*Aptenodytes patagonicus*) during summer at the Crozet Islands. *Ecology* 79:1905–1921.

Rattenborg, N. 2006. Do birds sleep in flight? *Naturwissenschaften* 93:413–425.

Robinson, S. A., and M. A. Hindell. 1996. Foraging ecology of gentoo penguins *Pygoscelis papua* at Macquarie Island during the period of chick care. *Ibis* 138 (4): 722–731.

Ropert-Coudert, Y., and A. K. A. Chiaradia. 2006. An exceptionally deep dive by a Little Penguin *Eudyptula minor. Marine Ornithology* 34:71–74.

Rutz, C., L. A. Bluff, N. Reed, J. Troscianko, J. Newton, R. Inger, A. Kacelnik, and S. Bearhop. 2010. The ecological significance of tool use in New Caledonian crows. *Science* 329:1523–1526.

Saraux, C., C. Le Bohec, J. M. Durant, V. A. Viblanc, M. Gauthier-Clerc, D. Beaune, Y.-H. Park, N. G. Yoccoz, N. C. Stenseth, and Y. Le Maho. 2011. Reliability of flipper-banded penguins as indicators of climate change. *Nature* 469:203–206.

Sato, K., K. Shiomi, G. Marshall, G. L. Kooyman, and P. J. Ponganis. 2011. Stroke rates and diving air volumes of emperor penguins: Implications for dive performance. *Journal of Experimental Biology* 214:2854–2863.

Scheffer, A., P. N. Trathan, and M. Collins. 2010. Foraging behaviour of King Penguins (*Aptenodytes patagonicus*) in relation to predictable mesoscale oceanographic features in the Polar Front Zone to the north of South Georgia. *Progress in Oceanography* 86:232–245.

Schmidt-Nielsen, K. 1983. *Animal Physiology: Adaptation and Environment.* Cambridge: Cambridge University Press.

Scholander, P. F. 1957. The wonderful net. *Scientific American* 196:96–107.

Schorger, A. W. 1947. The deep diving of the Loon and the Old-Squaw and its mechanism. *Wilson Bulletin* 59:151–159.

Scolaro, J. A., and A. M. Suburo. 1994. Timing and duration of foraging trips in magellanic penguins *Spheniscus magellanicus. Marine Ornithology* 22:231–235.

Seddon, P. J., and Y. V. Heezik. 1990. Diving depths of the Yellow-eyed penguin *Megadyptes antipodes. Emu* 90:53–57.

Shackleton, E. 1919. *South: The Story of Shackleton's Last Expedition, 1914–1917*. London: William Heibnemann.

Shaffer, S. A., H. Weimerskirch, D. Scott, D. Pinaud, D. R. Thompson, P. M. Sagar, H. Moller, G. A. Taylor, D. G. Foley, Y. Tremblay, and D. P. Costa. 2009. Spatiotemporal habitat use by breeding sooty shearwaters *Puffinus griseus. Marine Ecology, Progress Series* 391:209–220.

Simeone, A., M. Bernal, and J. Meza. 1999. Incidental mortality of Humboldt penguins *Spheniscus humboldti* in gill nets, central Chile. *Marine Ornithology* 27:157–161.

Simpson, G. G. 1975. Fossil penguins. In *The Biology of Penguins*, ed. B. Stonehouse, 19–56. London: Macmillan.

Simpson, G. G. 1976. *Penguins: Past and Present, Here and There.* New Haven, CT: Yale University Press.

Slack, K. E., C. M. Jones, T. Ando, G. L. Harrison, R. E. Fordyce, U. Aranson, and D. Penny. 2006. Early penguin fossils, plus mitochondrial genomes, calibrate avian evolution. *Molecular Biology and Evolution* 23:1144–1155.

Spellerberg, I. F. 1975. The predation of penguins. In *The Biology of Penguins*, ed. B. Stonehouse, 413–434. London: Macmillan.

Steinfurth, A., S. Emslie, W. P. Patterson, and S. Garthe. 2006. The diet of the Galapagos Penguin. *Journal of Ornithology* 147:257.

Stonehouse, B. 1969. Environmental temperatures of tertiary penguins. *Science* 163:673–675.

Stonehouse, B., and S. Stonehouse. 1963. The frigate bird *Fregata aethiops* of Ascension Island. *Ibis* 103b:409–422.

Suburo, A. M., and J. A. Scolaro. 1999. Environmental adaptations in the retina of the Magellanic Penguin: Photoreceptors and outer plexiform layer. *Waterbirds* 22:111–119.

Takahashi, A., M. J. Dunn, P. Trathan, K. Sato, Y. Naito, and J. P. Croxall. 2003. Foraging strategies of Chinstrap penguins at Signy Island, Antarctica: Importance of benthic feeding on Antarctic krill. *Marine Ecology Progress Series* 250:279–289.

Thomas, D. P., and G. F. Fregin. 1981. Cardiorespiratory and metabolic responses to treadmill exercise in the horse. *Journal of Applied Physiology* 50:864–868.

Thouzeau, C., Y. Le Maho, G. Froget, L. Sabatier, C. Le Bohec, J. A. Hoffmann, and P. Bulet. 2003. Spheniscins, avian ß-defensins in preserved stomach contents of the king penguin, *Aptenodytes patagonicus*. *Journal of Biological Chemistry* 278:51053–51058.

Thouzeau, C., G. Peters, C. Le Bohec, and Y. Le Maho. 2004. Adjustments of gastric pH, motility, and temperature during long-term preservation of stomach contents in free-ranging incubating king penguins. *Journal of Experimental Biology* 207:2715–24.

Todd, F. S. 2004. *Birds and Mammals of the Antarctic, Subantarctic and Falkland Islands*. 1st ed. Temecula, CA: Ibis Publishing Company.

Trathan, P. N. 2004. Image analysis of color aerial photography to estimate penguin population size. *Wildlife Society Bulletin* 33:232–243.

Trathan, P. N., P. T. Fretwell, and B. Stonehouse. 2011. First recorded loss of an emperor penguin colony in the recent period of Antarctic regional warming: Implications for other colonies. *PLoS ONE* 6:e14738.

Trathan, P. N., N. Ratcliffe, and E. A. Masden. 2012. Ecological drivers of change at South Georgia: The krill surplus, or climate variability. *Ecography* 35:1–11.

Tremblay, Y., T. R. Cook, and Y. Cherel. 2005. Time budget and diving behaviour of chick-rearing Crozet shags. *Canadian Journal of Zoology* 83:971–982.

Tremblay, Y., E. Guinard, and Y. Cherel. 1997. Maximum diving depths of northern rockhopper penguins (*Eudyptes chrysocome moseleyi*) at Amsterdam Island. *Polar Biology* 17:119–122.

Trivelpiece, W. Z., J. T. Hinke, A. K. Miller, C. S. Reiss, S. G. Trivelpiece, and G. M. Watters. 2011. Variability in krill biomass links harvesting and climate warming to penguin population changes in Antarctica. *Proceedings of the National Academy of Sciences* 108:7625–7628.

Trivelpiece, W. Z., S. G. Trivelpiece, and N. J. Volkman. 1987. Ecological segregation of Adelie, Gentoo, and Chinstrap penguins at King George Island, Antarctica. *Ecology* 68 (2): 351–361.

Underhill, L. G., P. A. Bartlett, L. Baumann, R. J. M. Crawford, B. M. Dyer, A. Gildenhuys, D. C. Nel, T. B. Oatley, M. Thornton, L. Upfold, et al. 1999. Mortality and survival of African penguins *Spheniscus demersus* involved in the Apollo Sea oil spill: An evaluation of rehabilitation efforts. *Ibis* 141:29–37.

Underhill, L. G., R. J. M. Crawford, A. C. Wolfaardt, P. A. Whittington, B. M. Dyer, T. M. Leshoro, M. Ruthenberg, L. Upfold, and J. Visagie. 2006. Regionally coherent trends in colonies of African penguins *Spheniscus demersus* in the Western Cape, South Africa, 1987 to 2005. *African Journal of Marine Science* 28:697–704.

Van Dam, R. P., and G. L. Kooyman. 2004. Latitudinal distribution of penguins, seals, and whales observed during a late autumn transect through the Ross Sea. *Antarctic Science* 16:313–318.

Vargas, H., C. Lougheed, and H. Snell. 2005. Population size and trends of the Galapagos penguin *Spheniscus mendiculus. Ibis* 147:367–374.

Viera, V. M., P. M. Nolan, S. D. Cote, P. Jouventin, and R. Groscolas. 2008. Is territory defence related to plumage ornaments in the king penguin *Aptenodytes patagonicus*? *Ethology* 114:146–153.

Vleck, C. M., M. F. Haussamann, and D. Vleck. 2003. The natural history of telomeres: Tools for aging animals and exploring the aging process. *Experimental Gerontology* 38:791–795.

Watanuki, Y., A. Kato, Y. Naito, G. Robertson, and S. Robinson. 1997. Diving and foraging behaviour of Adélie penguins in areas with and without fast sea-ice. *Polar Biology* 17:296–304.

Weimerskirch, H., P. Inchausti, C. Guinet, and C. Barbraud. 2003. Trends in bird and seal populations as indicators of a system shift in the Southern Ocean. *Antarctic Science* 15:249–56.

Wienecke, B., B. Raymond, and G. Robertson. 2010. Maiden journey of fledgling emperor penguins from the Mawson Coast, East Antarctica. *Marine Ecology Progress Series* 410:269–282.

Wienecke, B., G. Robertson, R. Kirkwood, and K. Lawton. 2007. Extreme dives by free-ranging emperor penguins. *Polar Biology* 30:133–142.

Williams, C. L., J. U. Meir, and P. J. Ponganis. 2011. What triggers the aerobic dive limit? Patterns of muscle oxygen depletion during dives of emperor penguins. *Journal of Experimental Biology* 214:1082–1112.

Williams, T. D. 1995. *The Penguins.* Bird Families of the World. Oxford: Oxford University Press.

Wilson, R. P. 1985. The jackass penguin (*Spheniscus demersus*) as a pelagic predator. *Marine Ecology Progress Series* 25:219–227.

Wilson, R. P., C. A. Bost, K. Puetz, J. B. Charrassin, B. M. Culik, and D. Adelung. 1997. Southern rockhopper penguin *Eudyptes chrysocome chrysocome* foraging at Possession Island. *Polar Biology* 17:323–329.

Wilson, R. P., B. Culik, D. Adelung, N. R. Coria, and H. J. Spairani. 1991. To slide or stride—when should Adélie penguins (*Pygoscelis adéliae*) toboggan. *Canadian Journal of Zoology* 69:221–225.

Wilson, R. P., P. G. Ryan, A. James, and M. P. T. Wilson. 1987. Conspicuous coloration may enhance prey capture in some piscivores. *Animal Behaviour* 35:1558–1600.

Yoda, K., K. Sato, Y. Niizuma, M. Kurita, C. A. Bost, Y. Le Maho, and Y. Naito. 1999. Precise monitoring of porpoising behaviour of Adélie penguins determined using acceleration data loggers. *Journal of Experimental Biology* 202:3121–3126.

Zachos, J., M. Pagani, L. Sloan, E. Thomas, and K. Billups. 2001. Trends, rhythms, and aberrations in global climate 65 Ma to present. *Science* 292:686–693.

Zavalaga, C. B., S. Benvenuti, A. Luigi Dall, and S. D. Emslie. 2007. Diving behavior of blue-footed boobies *Sula nebouxii* in northern Peru in relation to sex, body size, and prey type. *Marine Ecology Progress Series* 336:291–303.

Zavalaga, C., and J. Jahncke. 1997. Maximum dive depths of the Peruvian diving-petrel. *Condor* 99:1002–1004.

Index